Collins

ROYAL
OBSERVATORY
GREENWICH

2025 GUIDE
to the
NIGHT SKY
SOUTHERN HEMISPHERE

Radmila Topalovic, Storm Dunlop and Wil Tirion

Published by Collins
An imprint of HarperCollins Publishers
Westerhill Road, Bishopbriggs
Glasgow G64 2QT

HarperCollins Publishers
Macken House,
39/40 Mayor Street Upper,
Dublin 1, D01 C9W8, Ireland
www.harpercollins.co.uk

In association with
Royal Museums Greenwich, the group name for the National Maritime Museum,
Royal Observatory Greenwich, Queen's House and Cutty Sark 2015
www.rmg.co.uk

A catalogue record for this book is available from the British Library

ISBN 978-0-00-868815-8

10 9 8 7 6 5 4 3 2 1

Printed in Malaysia

If you would like to comment on any aspect of this book, please contact us at the above address or online.
e-mail: collins.reference@harpercollins.co.uk

Contents

Introduction

The aim of this Guide is to help people find their way around the night sky at any time of the year, by showing how the stars that are visible change from month to month and highlighting various events that occur throughout the year. The objects and events described may be observed with the naked eye, or nothing more complicated than a pair of binoculars.

The conditions for observing naturally vary over the course of the year. During the summer, twilight may persist throughout the night and make it difficult to see the faintest stars. There are three recognized stages of twilight: civil twilight, when the Sun is less than 6° below the horizon; nautical twilight, when the Sun is between 6° and 12° below the horizon; and astronomical twilight, when the Sun is between 12° and 18° below the horizon. Full darkness occurs only when the Sun is more than 18° below the horizon. During nautical twilight, only the very brightest (navigation) stars are visible. During astronomical twilight, the faintest stars visible to the naked eye may be seen directly overhead, but are lost at lower altitudes. At Sydney, full darkness persists for about six hours at mid-summer. Even at

Christchurch, NZ (not shown), full darkness lasts about four hours. By contrast, nautical twilight persists as far south as Cape Horn at mid-summer, so only the very brightest stars are visible.

Another factor that affects the visibility of objects is the amount of moonlight in the sky. At Full Moon, it may be very difficult to see some of the fainter stars and objects, and even when the Moon is at a smaller phase it may seriously interfere with visibility if it is near the stars or planets in which you are interested. A full lunar calendar is given for each month and may be used to see when nights are likely to be darkest and best for observation.

The celestial sphere

All the objects in the sky (including the Sun, Moon and stars) appear to lie at some indeterminate distance on a large sphere, centred on the Earth. This *celestial sphere* has various reference points and features that are related to those of the Earth. If the Earth's rotational axis is extended, for example, it points to the North and South Celestial Poles, which are thus in line with the North and South Poles on Earth. Similarly, the *celestial equator*

The duration of twilight throughout the year at Sydney and Cape Horn.

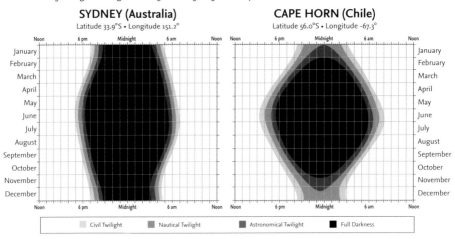

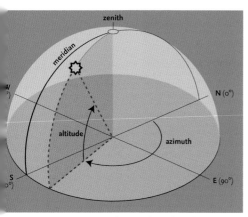

Measuring altitude and azimuth on the celestial sphere.

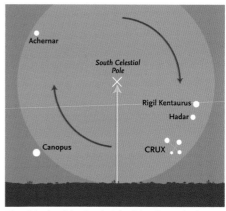

The altitude of the South Celestial Pole equals the observer's latitude.

lies in the same plane as the Earth's equator, and divides the sky into northern and southern hemispheres. Because this Guide is written for use in the southern hemisphere, the area of the sky that it describes includes the whole of the southern celestial hemisphere and those portions of the northern that become visible at different times of the year. Stars in the far north, however, remain invisible throughout the year, and are not included.

It is useful to know some of the special terms for various parts of the sky. As seen by an observer, half of the celestial sphere is invisible at any point in time; these objects will be below the horizon. The point directly overhead is known as the **zenith**, and the (invisible) one below one's feet as the **nadir**. The line running from the north point on the horizon, up through the zenith and then down to the south point is the **meridian**. This is an important invisible line in the sky, because objects are highest in the sky, and thus easiest to see, when they cross the meridian. Objects are said to **transit**, when they cross this line in the sky.

In this book, reference is frequently made in the text and in the diagrams to the standard compass points around the horizon. The position of any object in the sky may be described by its **altitude** (measured in degrees

above the horizon), and its **azimuth** (measured in degrees from north 0°, through east 90°, south 180° and west 270°). Experienced amateurs and professional astronomers also use another system of specifying locations on the celestial sphere, but that need not concern us here, where the simpler method will suffice.

The celestial sphere appears to rotate about an invisible axis, running between the North and South Celestial Poles. The location (i.e. the altitude) of the Celestial Poles depends entirely on the observer's position on Earth or, more specifically, their latitude. The charts in this book are produced for the latitude of 35°S, so the South Celestial Pole (SCP) is 35° above the southern horizon. The fact that the SCP is fixed relative to the horizon means that all the stars within 35° of the pole are always above the horizon and may, therefore, always be seen at night, regardless of the time of year. The southern circumpolar region is an ideal place to begin learning the sky, and ways to identify the circumpolar stars and constellations will be described shortly.

The ecliptic and the zodiac

Another important line on the celestial sphere is the Sun's apparent path against the background stars – in reality the result of the Earth's orbit around the Sun. This is known

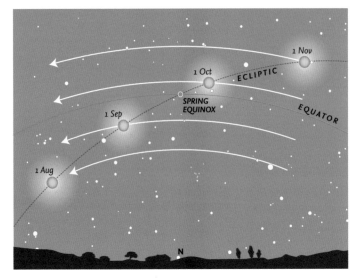

The Sun crossing the celestial equator at the September equinox (spring equinox in the southern hemisphere.).

as the *ecliptic*. The point where the Sun, apparently moving along the ecliptic, crosses the celestial equator from south to north is known as the (southern) autumn equinox, which occurs on 20 or 21 March. At this time and at the (southern) spring equinox, on 22 or 23 September, when the Sun crosses the celestial equator from north to south, day and night are almost exactly equal in length. (There is a slight difference, but that need not concern us here.) The March equinox is currently located in the constellation of Pisces, and is important in astronomy because it defines the zero point for a system of celestial coordinates, which is, however, not used in this Guide.

The Moon and planets are to be found in a band of sky that extends 8° on either side of the ecliptic. This is because the orbits of the Moon and planets are inclined at various angles to the ecliptic (i.e. to the plane of the Earth's orbit). This band of sky is known as the zodiac and, when originally devised, consisted of twelve **constellations**, all of which were considered to be exactly 30° wide. When the constellation boundaries were formally established by the International Astronomical Union in 1930, the exact extent of most

constellations was altered and, nowadays, the ecliptic passes through thirteen constellations. Because of the boundary changes, the Moon and planets may actually pass through several other constellations that are adjacent to the original twelve.

The constellations

Since ancient times, the celestial sphere has been divided into various constellations, most dating back to antiquity and usually associated with certain myths or legendary people and animals. Nowadays, the boundaries of the constellations have been fixed by international agreement and their names (in Latin) are largely derived from Greek or Roman originals. Some of the names of the most prominent stars are of Greek or Roman origin, but many are derived from Arabic names. Many bright stars have no individual names and, for many years, stars were identified by terms such as 'the star in Hercules' right foot'. A more sensible scheme was introduced by the German astronomer Johannes Bayer in the early seventeenth century. Following his scheme – which is still used today – most of the brightest stars are identified by a Greek letter followed by the genitive form of the constellation's Latin name. An example is the Pole Star, also known as Polaris and α Ursae Minoris (abbreviated α UMi). The Greek alphabet is shown on page 110 and a list of all the constellations that may be seen from latitude 35°S, together with abbreviations, their genitive forms and English names is on page 109. Other naming schemes exist for fainter stars, but are not used in this book.

Asterisms

Apart from the constellations (88 of which cover the whole sky), certain groups of stars, which may form a part of a larger constellation or cross several constellations, are readily recognizable and have been given individual names. These groups are known as *asterisms*, and the most famous (and well-known to northern observers) is the 'Plough', the common name for the seven brightest stars in the constellation of Ursa Major, the Great Bear. The names and details of some asterisms mentioned in this book are given in the list on page 110.

Magnitudes

The brightness of a star, planet or other body is frequently given in magnitudes (mag.). This is a mathematically defined scale where larger numbers indicate a fainter object. The scale extends beyond the zero point to negative numbers for very bright objects. (Sirius, the brightest star in the sky is mag. -1.4.) Most observers are able to see stars as faint as about mag. 6, under very clear skies.

The Moon

Although the daily rotation of the Earth carries the sky from east to west, the Moon gradually moves eastwards relative to the stars by approximately its diameter (about half a degree) in an hour. Normally, in its orbit around the Earth, the Moon passes above or below the direct line between Earth and Sun (at New Moon) or outside the area obscured by the Earth's shadow (at Full Moon). Occasionally, however, the three bodies are more-or-less perfectly aligned to give an *eclipse*: a solar eclipse at New Moon or a lunar

eclipse at Full Moon. Depending on the exact circumstances, a solar eclipse may be merely partial (when the Moon does not cover the whole of the Sun's disk); annular (when the Moon is too far from Earth in its orbit to appear large enough to hide the whole of the Sun); or total. Total and annular eclipses are visible from very restricted areas of the Earth, but partial eclipses are normally visible over a wider area.

Somewhat similarly, at a lunar eclipse, the Moon may pass through the outer zone of the Earth's shadow, the **penumbra** (in a penumbral eclipse, which is not generally perceptible to the naked eye), so that just part of the Moon is within the darkest part of the Earth's shadow, the **umbra** (in a partial eclipse); or completely within the umbra (in a total eclipse). Unlike solar eclipses, lunar eclipses are visible from large areas of the Earth.

Occasionally, as it moves across the sky, the Moon passes between the Earth and individual planets or distant stars, giving rise to an **occultation**. As with solar eclipses, such occultations are visible from restricted areas of the world.

The planets

The planets are always moving against the background stars, therefore they are treated in some detail in the monthly pages and information is given when they are close to other planets, the Moon or any of five bright stars that lie near the ecliptic. Such events are known as **appulses** or, more frequently, as conjunctions. (There are technical differences in the way these terms are defined – and should be used – in astronomy, but these need not concern us here.) The positions of the planets are shown for every month on a special chart of the ecliptic

The term conjunction is also used when a planet is either directly behind or in front of the Sun, as seen from Earth. (Under normal circumstances it will then be invisible.) The conditions of most favourable visibility depend on whether the planet is one of the two known

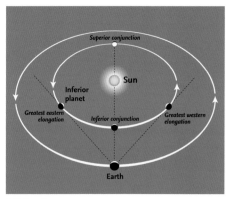

Inferior planet.

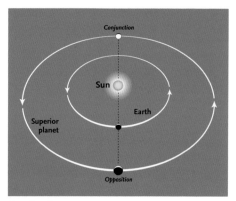

Superior planet.

as **inferior planets** (Mercury and Venus) or one of the three **superior planets** (Mars, Jupiter and Saturn) that are covered in detail. (Some details of the fainter superior planets, Uranus and Neptune, are included in this Guide, and special charts for both are given on page 25.)

The inferior planets are most readily seen at eastern or western **elongation**, when their angular distance from the Sun (separation from the Sun in degrees) is greatest. For superior planets, they are best seen at **opposition**, when they are directly opposite the Sun in the sky, and cross the meridian at local midnight.

It is often useful to be able to estimate angles on the sky, and approximate values may be obtained by holding one hand at arm's length. The various angles are shown in the diagram, together with the separations of the various stars in and around Orion.

Meteors

At some time or other, nearly everyone has seen a **meteor** – a 'shooting star' – as it flashed across the sky. The particles that cause meteors – known technically as 'meteoroids' – range in size from that of a grain of sand (or even smaller) to the size of a pea. On any night of the year there are occasional meteors, known as **sporadics**, that may travel in any direction. These occur at a rate that is normally between three and eight in an hour. Far more important, however, are **meteor showers**, which

occur at fixed periods of the year, when the Earth encounters a trail of particles left behind by a comet or, very occasionally, by a minor planet (asteroid). Meteors always appear to diverge from a single point on the sky, known as the **radiant**, and the radiants of major showers are shown on the charts. Meteors that come from a circular area 8° in diameter around the radiant are classed as belonging to the particular shower. All others that do not come from that area are sporadics (or, occasionally from another shower that is active at the same time). A list of the major meteor showers is given on page 31.

Although the positions of the various shower radiants are shown on the charts, looking directly at the radiant is not the most effective way of seeing meteors. They are most likely to be noticed if one is looking about 40–45° away from the radiant position. (This is approximately two hand-spans as shown in the diagram for measuring angles.)

Other objects

Certain other objects may be seen with the naked eye under good conditions. Some were given names in antiquity – Praesepe is one example – but many are known by what are called 'Messier numbers', the numbers in a catalogue of nebulous objects compiled by Charles Messier in the late eighteenth century. Some, such as the Andromeda Galaxy, M31, and the Orion Nebula, M42, may be seen

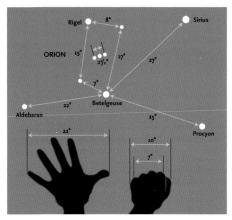

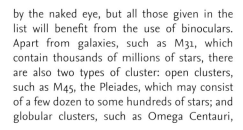

Measuring angles in the sky.

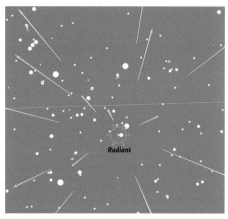

Meteor shower (showing the Geminids radiant).

by the naked eye, but all those given in the list will benefit from the use of binoculars. Apart from galaxies, such as M31, which contain thousands of millions of stars, there are also two types of cluster: open clusters, such as M45, the Pleiades, which may consist of a few dozen to some hundreds of stars; and globular clusters, such as Omega Centauri, which are spherical concentrations of many thousands of stars. One or two gaseous nebulae (emission nebulae), consisting of gas illuminated by stars within them, are also visible. The Orion Nebula, M42, is one, and is illuminated by the group of four stars, known as the Trapezium, which may be seen within it by using a good pair of binoculars.

Some interesting objects.

Messier / NGC	Name	Type	Constellation	Maps (months)
—	47 Tucanae	globular cluster	Tucana	All year
—	Hyades	open cluster	Taurus	Oct–Feb
—	Melotte 111 (Coma Cluster)	open cluster	Coma Berenices	Mar–Jun
M3	—	globular cluster	Canes Venatici	Mar–Jul
M4	—	globular cluster	Scorpius	Mar–Sep
M8	Lagoon Nebula	gaseous nebula	Sagittarius	Apr–Oct
M11	Wild Duck Cluster	open cluster	Scutum	May–Oct
M13	Hercules Cluster	globular cluster	Hercules	May–Aug
M15	—	globular cluster	Pegasus	Jul–Nov
M20	Trifid Nebula	gaseous nebula	Sagittarius	Apr–Oct
M22	—	globular cluster	Sagittarius	Apr–Oct
M27	Dumbbell Nebula	planetary nebula	Vulpecula	Jun–Oct
M31	Andromeda Galaxy	galaxy	Andromeda	Sep–Dec
M35	—	open cluster	Gemini	Nov–Mar
M42	Orion Nebula	gaseous nebula	Orion	Oct–Apr
M44	Praesepe	open cluster	Cancer	Dec–Apr
M45	Pleiades	open cluster	Taurus	Oct–Feb
M57	Ring Nebula	planetary nebula	Lyra	Jun–Sep
M67	—	open cluster	Cancer	Dec–May
IC 2602	Southern Pleiades	open cluster	Carina	All year
NGC 2070	Tarantula Nebula	emission nebula	Dorado (LMC)	All year
NGC 3242	Ghost of Jupiter	planetary nebula	Hydra	Jan–Jun
NGC 3372	Eta Carinae Nebula	gaseous nebula	Carina	All year
NGC 4755	Jewel Box	open cluster	Crux	All year
NGC 5139	Omega Centauri	globular cluster	Centaurus	Jan–Sep

The Southern Circumpolar Constellations

The southern circumpolar constellations are the key to to starting to identify the constellations. For anyone in the southern hemisphere, they are visible at any time of the year, and nearly everyone is familiar with the striking pattern of four stars that make up the constelllation of **Crux** (the Southern Cross), and also the two nearby bright stars **Rigil Kentaurus** and **Hadar** (α and β Centauri, respectively). This pattern of stars is visible throughout the year for most observers, although for observers farther north, the stars may become difficult to see - low on the horizon in the southern spring, especially in the months of October and November.

Crux

The distinctive shape of the constellation of **Crux** is usually easy to identify, although some people (especially northerners unused to the southern sky) may wrongly identify the slightly larger False Cross, formed by the stars **Aspidiske** and **Avior** (ι and ε Carinae respectively) plus **Markeb** and **Alsephina** (κ and δ Velorum). The dark patch of the Coalsack (a dark cloud of obscuring dust) is readily visible on the eastern side of Crux between **Acrux** and **Mimosa** (α and β Crucis).

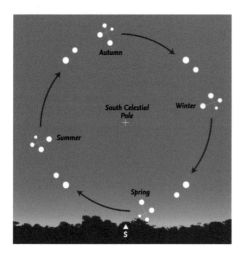

A line through **Gacrux** (γ Crucis) and **Acrux** (α Crucis) points approximately in the direction of the South Celestial Pole, crossing the faint constellations of **Musca** and **Chamaeleon**. Although there is no bright star close to the South Celestial Pole (and even the constellation in which it lies, **Octans**, is faint), an idea of its location helps to identify the region of sky that is always visible. (The altitude of the South Celestial Pole is equal to the observer's latitude south of the equator.) The basic triangular shape of Octans itself is best found by extending a line from **Peacock** (α Pavonis) through β Pavonis, by about the same distance as that between the stars.

Centaurus

Although Crux is a distinctive shape, **Centaurus** is a large, rather straggling constellation, with one notable object: the giant, bright globular cluster **Omega Centauri** (the brightest globular in the sky), which lies towards the north, on the line from **Hadar** (β Centauri) through ε Centauri. Rather than locating the South Celestial Pole from Crux, a better indication is the line, at right angles to the line between Hadar and Rigil Kentaurus, that passes across the brightest star in **Circinus** (α).

Carina

Apart from the two stars that form part of the False Cross, the constellation of Carina is, like Centaurus, a large, sprawling constellation. It contains one striking open cluster, the **Southern Pleiades**, and a remarkable emission nebula, the **Eta Carinae Nebula**. The second brightest star in the sky (after Sirius) is **Canopus**, α Carinae, which lies far away to the west.

The Magellanic Clouds

On the opposite side of the South Celestial Pole to Crux and Centaurus lie the two Magellanic Clouds. The **Small Magellanic Cloud** (SMC) lies to one side of the relatively inconspicuous, triangular constellation of **Hydrus**, but is actually within the constellation of **Tucana**.

The stars and constellations inside the circle are always above the horizon, seen from latitude 35°S.

Nearby is another bright globular cluster, **47 Tucanae**. Hydrus itself is also easily identified from the star **Achernar**, α Eridani, the rather isolated brilliant star at the southern end of **Eridanus**, which wanders a long way south, having begun at the foot of Orion.

The **Large Magellanic Cloud** (LMC) lies within the faint constellation of **Dorado**. Not only is it a large satellite galaxy of the Milky Way Galaxy, but it contains the large, readily visible, **Tarantula Nebula**, an emission nebula that is a major star-forming region.

The Summer Constellations

The summer sky is dominated by the constellation of **Orion**, with its three 'Belt' stars, as well as by the three bright stars: **Betelgeuse** (α Orionis), **Sirius** (α Canis Majoris), and **Procyon** (α Canis Minoris), which together form the prominent (Southern) 'Summer Triangle'. To the west of Orion is **Taurus**, with orange **Aldebaran** (α Tauri). A line from **Bellatrix** (γ Orionis) through Aldebaran points to the striking **Pleiades** cluster (M45). (The line of three 'Belt' stars indicates the same general path and, in the other direction, points towards Sirius.) A line from **Mintaka** (the westernmost star of the 'Belt') through brilliant **Rigel** (β Orionis) points in the direction of **Achernar** (α Eridani). One from **Alnitak** (at the other end of the 'Belt') through **Saiph** (κ Orionis) indicates the direction of **Canopus** (α Carinae), the second brightest star in the whole sky. A line from Betelgeuse through Sirius points in the general direction of the 'False Cross' in **Vela** and **Carina**.

In the opposite (northern) side of Orion, a line from **Meissa** (λ Orionis) through **Elnath** (β Tauri) leads to **Capella** (α Aurigae), and one from **Betelgeuse** through **Alhena** (γ Geminorum) points to **Pollux** (β Geminorum), the southernmost of the distinctive Castor/Pollux pair. A line across Orion from Bellatrix to Betelgeuse, if extended greatly, indicates the general direction of **Regulus** (α Leonis).

Six of the bright stars that have been mentioned, in different constellations: Capella, Aldebaran, Rigel, Sirius, Procyon and Pollux form what, for northern observers, is sometimes known as the 'Winter Hexagon'. Pollux is accompanied to the northwest by the slightly fainter star of **Castor** (α Geminorum), the second 'Twin'.

Several of the stars in this region of the sky show distinctive tints: Betelgeuse (α Orionis) is reddish, Aldebaran (α Tauri) is orange and Rigel (β Orionis) is blue-white.

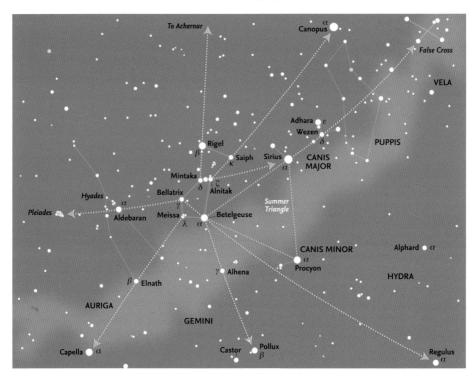

The Autumn Constellations

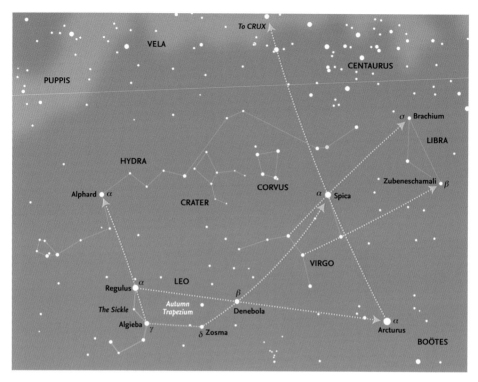

During the autumn season, the principal constellation is the zodiacal constellation of **Leo**, whose principal star **Regulus** (α Leonis) is one of the five bright stars that may occasionally be occulted by the Moon. Regulus forms the 'dot' of the upside-down 'question mark' of stars, known as the 'Sickle'. The body of Leo itself is sometimes known as the 'Trapezium' (here, the 'Autumn Trapezium'), and its sides may be used to find other objects in the sky.

A line extended from **Algieba** (γ Leonis), the second star in the 'Sickle', through Regulus points to **Alphard** (α Hydrae) the brightest star in **Hydra**, the largest of the 88 constellations, which sprawls a long way across the sky from its distinctively shaped 'head' of stars, southwest of Regulus. A line from Regulus through **Denebola** (β Leonis) at the other end of the constellation of Leo points towards **Arcturus** (α Boötis), the brightest star in the northern hemisphere of the sky. Another line, down the 'back' of Leo, from **Zosma** (δ Leonis) through Denebola points in the general direction of **Spica** (α Virginis) the principal star in the constellation of **Virgo**.

Although Spica is the brightest star in the zodiacal constellation of Virgo, the remainder of the constellation is not remarkable, consisting of a rough quadrilateral of stars with fainter lines extending outwards. The two sides of the main quadrilateral, if extended, point eastwards towards the two principal stars, **Zubeneschamali** and **Brachium** (β and σ Librae, respectively), of the small zodiacal constellation of **Libra**. The two small constellations of **Corvus** and **Crater** lie between Virgo and the long stretch of Hydra. A line from Arcturus through Spica, if extended far to the south across the sky, points to the distinctive constellation of **Crux**.

The Winter Constellations

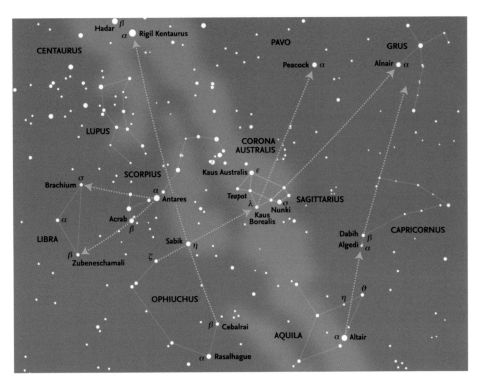

During the winter season, the Milky Way runs right across the sky, and the distinctive constellation of **Scorpius** is clearly visible, with bright red **Antares** (α Scorpii) and the constellation's 'sting' trailing behind it. Scorpius used to be a much larger constellation until the small zodiacal constellation of **Libra** was formed from some of its stars. The two sides of the 'fan' of stars immediately east of Antares, if extended, point to the two principal stars of Libra, **Zubeneschamali** and **Brachium** (β and σ Librae, respectively) on the southern and northern sides of the constellation.

North of Scorpius lies the large constellation of **Ophiuchus**. The ecliptic runs through the southern portion of Ophiuchus, and the Sun spends far more time in that constellation than it does in neighbouring Scorpius. A line from **Cebalrai** (β Ophiuchi) in the north of the constellation, through **Sabik** (η Ophiuchi), if extended right across

Scorpius, carries you to the two bright stars of **Centaurus**, **Rigil Kentaurus** and **Hadar** (α and β Centauri, respectively). Between Libra and Centaurus lies the straggling constellation of **Lupus** with no distinct pattern of stars.

A line from the fainter star ζ Ophiuchi through **Sabik** points towards **Kaus Borealis** (λ Sagittarii) the 'lid' of the 'Teapot' of **Sagittarius**. From there two lines radiate (on the west) across **Corona Australis** towards **Peacock** (α Pavonis) and (on the east) **Alnair** (α Gruis) in the small constellation of **Grus**.

Also north of **Scorpius** is the constellation of **Aquila**, with its brightest star, **Altair** (α Aquilae). The diamond shape of the 'wings' of Aquila may be used to locate **Algedi** (α Capricorni), a prominent double star in **Capricornus**. A line through **Dabih** (β Capricorni) and the western side of that basically triangular constellation, when extended, also points in the direction of the constellation of **Grus**.

The Spring Constellations

In spring, the constellation of **Pegasus** and the Great Square of Pegasus are prominent in the north. One star in the Square, **Alpheratz**, at the northwestern corner, actually belongs to the neighbouring constellation of **Andromeda**. A diagonal line across the Square from Alpheratz (α Andromedae) through **Markab** (α Pegasi) points to **Sadalmelik** (α Aquarii), just to the west of the 'Y-shaped' asterism of the 'Water Jar' in the zodiacal constellation of **Aquarius**. Further extended, that line carries the observer to the centre of the triangular constellation of **Capricornus**.

The constellation of **Pisces** consists of two lines of stars running to the south and east of the Great Square. In mythological representations, the two fish are linked by ribbons, tied together at **Alrescha** (α Piscium). The distinctive asterism of **The Circlet** lies at the western end of the southern line of stars. The two north-south lines of the Great Square may

be used as guides to other constellations. The western line from **Scheat** (β Pegasi) through Markab, if extended about three times, crosses Aquarius and points to the bright, fairly isolated star, **Fomalhaut** (α Piscis Austrini) in the constellation of **Piscis Austrinus** and then onward to **Tiaki** (β Gruis) in the centre of a cross of stars within **Grus**. The eastern line, from Alpheratz through **Algenib** (γ Pegasi), extended about three-and-a-half times, indicates **Ankaa** (α Phoenicis) in the constellation of **Phoenix**.

The line between Markab and Algenib may be extended as an arc to lead to **Menkar** (α Ceti) in the 'tail' of Cetus. A line from Menkar through **Diphda** (β Ceti) also points to Fomalhaut in Piscis Austrinus. Diphda and Fomalhaut form the base of an almost perfect, large isoceles triangle, with **Achernar** (α Eridani) at the other apex. **Eridanus** itself is a long, trailing constellation that begins far to the north, near **Rigel** in **Orion**.

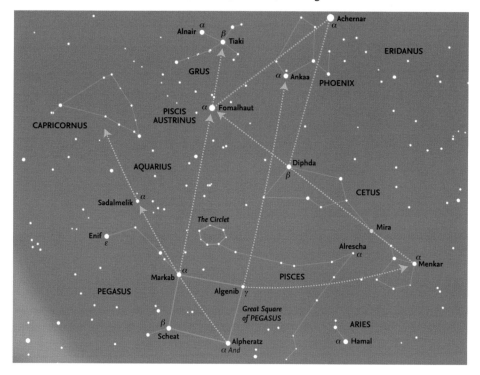

The Moon

The monthly pages include diagrams showing the phase of the Moon for every day of the month, and also indicate the day in the **lunation** (or **age** of the Moon), which begins at New Moon. Although the main features of the surface – the light highlands and the dark maria (seas) – may be seen with the naked eye, far more features may be detected with the use of binoculars or any telescope. The many craters are best seen when they are close to the **terminator** (the boundary between the illuminated and the non-illuminated areas of the surface), when the Sun rises or sets over any particular region of the Moon and the crater walls or central peaks cast strong shadows. Most features become difficult to see at Full Moon, although this is the best time to see the bright ray systems surrounding certain craters. Accompanying the Moon map on the following pages is a list of prominent features, including the days in the lunation when they are normally close to the terminator and thus easiest to see. A few bright features

such as Linné and Proclus, visible when well illuminated, are also listed. One feature, Rupes Recta (the Straight Wall) is readily visible only when it casts a shadow with light from the east, appearing as a light line when illuminated from the opposite direction.

The dates of visibility vary slightly through the effects of **libration**. Because the Moon's orbit is inclined to the Earth's equator and also because it moves in an ellipse, the Moon appears to rock slightly from side to side (and nod up and down). Areas near the **limb** (the edge of the Moon) may vary considerably in their location and visibility. (This is easily noticeable with Mare Crisium and the craters Tycho and Plato.) Another effect is that at crescent phases before and after New Moon, the normally non-illuminated portion of the Moon receives a certain amount of light, reflected from the Earth. This **Earthshine** may enable certain bright features (such as Aristarchus, Kepler and Copernicus) to be detected even though they are not illuminated by sunlight.

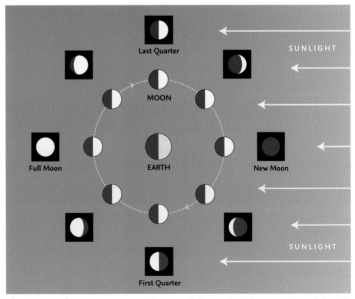

The Moon phases. *During its orbit around the Earth, different portions of the side of the Moon that faces us are illuminated by the Sun.*

The Moon at First Quarter (south is up).

Map of the Moon

The numbers indicate the day or days (the age) of the Moon when features are usually best visible.

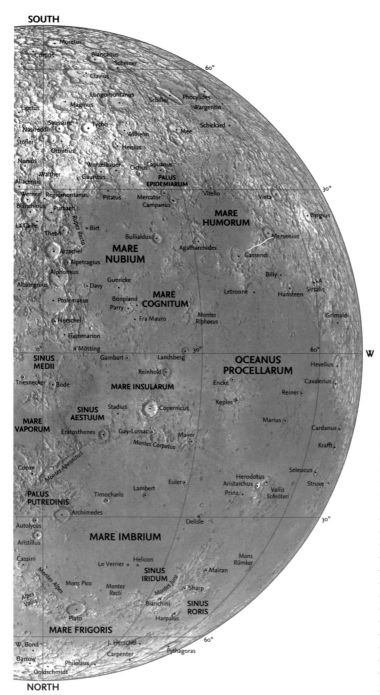

SOUTH

Moretus
Curtius
Blancanus
Scheiner
60°
Clavius
Longomontanus
Schiller
Phocylides
Licetus
Maginus
Wargentin
Saussure
Tycho
Wilhelm
Schickard
Nasireddin
Mee
Stöfler
Orontius
Heisius
Nonius
Wurzelbauer
Cichus
Capuanus
Walther
Gauricus
PALUS
EPIDEMIARUM
Aliacensis
Werner
Regiomontanus
Pitatus
Mercator
Vitello
30°
Blanchinus
Purbach
Campanus
Vieta
La Caille
Thebit
Birt
MARE
HUMORUM
Byrgius
Rupes Recta
Bullialdus
Mersenius
Arzachel
MARE
NUBIUM
Agatharchides
Alpetragius
Gassendi
Alphonsus
Guericke
Billy
Albategnius
Davy
MARE
COGNITUM
Letronne
Hansteen
Sirsalis
Ptolemaeus
Bonpland
Parry
A
Herschel
Fra Mauro
Montes
Riphaeus
Grimaldi
Flammarion
Mösting
30°
60°
W
SINUS
MEDII
Gambart
Landsberg
OCEANUS
PROCELLARUM
Hevelius
Triesnecker
Bode
Reinhold
Cavalerius
MARE INSULARUM
Encke
Reiner
MARE
VAPORUM
SINUS
AESTUUM
Stadius
Copernicus
Kepler
Marius
Eratosthenes
Gay-Lussac
Mayer
Cardanus
Montes Carpatus
Krafft
Cocon
Herodotus
Seleucus
Aristarchus
Struve
PALUS
PUTREDINIS
Timocharis
Lambert
Euler
Prinz
Vallis
Schröteri
Archimedes
Delisle
30°
Autolycus
Aristillus
MARE IMBRIUM
Cassini
Le Verrier
Helicon
Mons
Rümker
Montes Alpes
Mons Pico
Montes
Recti
SINUS
IRIDUM
Mairan
Alpes
Vallis
Bianchini
Sharp
SINUS
RORIS
Plato
Harpalus
MARE FRIGORIS
60°
W. Bond
J. Herschel
Pythagoras
Barrow
Carpenter
Philolaus
Goldschmidt
NORTH

Eclipses in 2025

Lunar eclipses

There are two total lunar eclipses in 2025, the first takes place on 14 March and it will be visible from Western Europe as well as North and South America, West Africa and the Pacific Ocean. The second eclipse occurs on 7 September, this will be visible from Europe, Africa, Asia and Australia. During a total lunar eclipse the Moon moves eastward into the Earth's penumbra, then moves behind the Earth into the darkest region of the Earth's shadow – the umbra – eventually creeping back out into the penumbra and into the sunlight. On 14 March maximum eclipse occurs at 06:59, the Moon will spend a total of 1 hour and 6 minutes immersed in the darkest shadow of the Earth. During this time the Moon is still visible; however, it will appear a reddish-brown colour, this is due to refracted sunlight which passes through the Earth's atmosphere and bends towards the lunar surface. Bluer hues are scattered outwards by the Earth's atmosphere, leaving behind redder light which eventually reaches the Moon. On the evening of 7 September the time of maximum eclipse will be 18:12; the umbral eclipse will last 1 hour 22 minutes.

Solar eclipses

There are two partial solar eclipses in 2025: the first takes place on 29 March with the maximum occurring at 11:03 from London and a duration of 1 hour and 53 minutes. This eclipse will be visible from Europe, northwest Africa and northern Russia. The second partial eclipse occurs on 21 September; this will be visible only from the Southern Hemisphere.

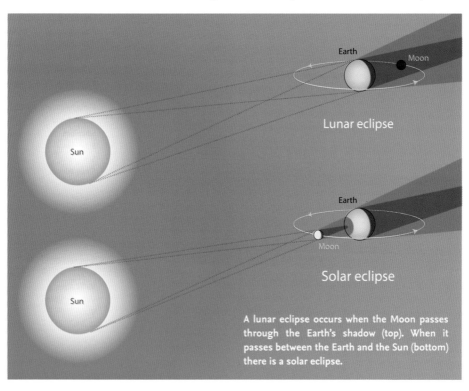

A lunar eclipse occurs when the Moon passes through the Earth's shadow (top). When it passes between the Earth and the Sun (bottom) there is a solar eclipse.

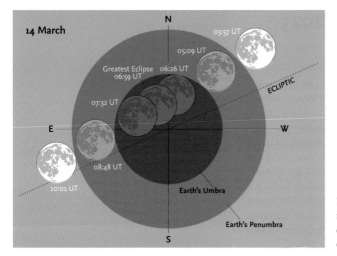

14 March

03:57 UT
05:09 UT
Greatest Eclipse 06:26 UT
06:59 UT
ECLIPTIC
07:32 UT
E — W
08:48 UT
10:01 UT
Earth's Umbra
Earth's Penumbra
N
S

The path of the Moon as it passes through the shadow of the Earth on 14 March. Maximum eclipse occurs at 06:59 UT.

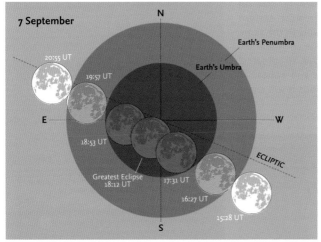

7 September

N
20:55 UT
19:57 UT
Earth's Penumbra
Earth's Umbra
E — W
18:53 UT
ECLIPTIC
Greatest Eclipse 17:31 UT
18:12 UT
16:27 UT
15:28 UT
S

The path of the Moon as it passes through the shadow of the Earth on 7 September. Maximum eclipse occurs at 18:12 UT.

A sequence of five images of the Moon, taken by Akira Fuji, showing the progression of the Moon (from right to left: west to east) through the Earth's shadow.

The Planets in 2025

Mercury and Venus

Mercury reaches greatest eastern elongation three times during 2025 – it will be seen in the early evening on 8 March (with Venus nearby), 4 July and 29 October, although it will be just above the horizon at sunset on this day (with Mars in the same region of the sky). Mercury switches to an appearance at dawn when it is at greatest western elongation on 21 April, 19 August and 7 December, although in April it will be positioned at a low altitude with Venus and Saturn close by. Its magnitude oscillates throughout the year – it will be brightest in May (-2.3) and faintest in November (6.1).

    Venus appears as the evening star on 10 January when it is at greatest eastern elongation. Saturn will be visible in the same area and both planets will lie in Aquarius. On 1 June Venus will move to greatest western elongation in Pisces and appear in the morning. On both occasions it will shine brightly at mag. -4.4.

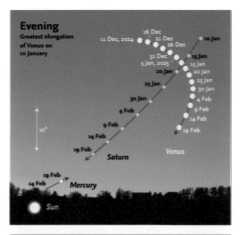

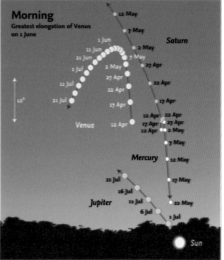

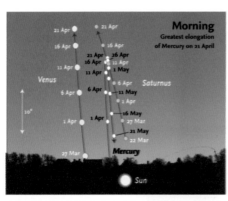

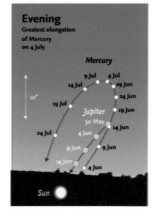

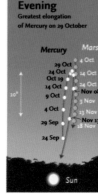

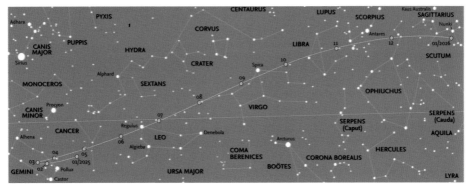

The path of Mars in 2025. In the last months it is caught up by the Sun and disappears in the evening twilight. South is up.

Mars

Mars will reach perigee on 12 January and opposition takes place 4 days later due to its elliptical orbit. It will reach a peak magnitude of -1.4 in **Gemini**. Mars does not come to opposition every year, the last opposition took place in December 2022. Instead it often remains in the sky for many months at a time, moving slowly along the ecliptic. Oppositions occur during a period of retrograde motion, when the planet appears to move westwards against the pattern of distant stars. Mars began retrograde motion on 6 December 2024 and it will end its westward motion on 24 February, moving eastward until 10 January 2027 when it will appear to move backwards again as Earth overtakes it in its orbit around the Sun.

As Mars moves away from Earth, its apparent size in the sky will decrease and it will dim to a magnitude of 1.6 in August, slowly appearing brighter towards the end of the year as it reaches magnitude 1.0. Mars starts in **Cancer**, moves into **Gemini** then back to Cancer in April, it will pass through **Leo** and **Virgo** in the summer, setting earlier in the

evening and appearing closer to the Sun as Autumn approaches. Mars will be in **Libra** in October and then it will move through **Scorpius** and **Ophiuchius** all before December.

Because of its eccentric orbit, which carries it at very differing distances from the Sun (and Earth), not all oppositions of Mars are equally favourable for observation. The relative positions of Mars and the Earth are shown here. It will be seen that the opposition of 2018 was very close and thus favourable for observation, and that of 2020 was also reasonably good. By comparison, opposition in 2027 will be at a far greater distance, so the planet will appear much smaller.

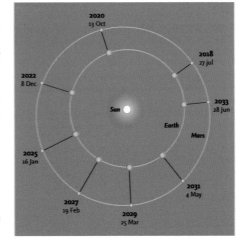

The oppositions of Mars between 2018 and 2033. As the illustration shows, there is an opposition on 16 January 2025.

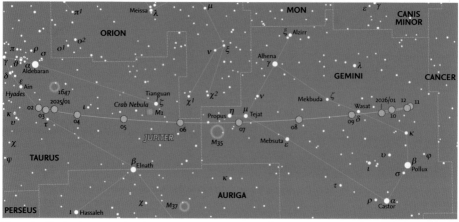

The path of Jupiter in 2025. Jupiter does not come to opposition until 10 January, 2026. Background stars are shown down to magnitude 6.5. South is up on all maps.

Jupiter and Saturn

Jupiter reaches opposition in early December 2024 and begins 2025 in **Taurus** at a magnitude of -2.7, it ends retrograde motion on 4 February, 2025. It moves into **Gemini** in June, dimming to mag. -1.9 and too close to the Sun to be seen. Jupiter will enter retrograde motion again on 11 November, brightening to -2.7 in December.

Jupiter's four large satellites are readily visible in binoculars. Not all four are visible all the time, they are sometimes hidden behind the planet or invisible in front of it. Io, the closest to Jupiter, orbits in just under 1.8 days,

and **Callisto**, the farthest away, takes about 16.7 days. In between are **Europa** (c. 3.6 days) and the largest, **Ganymede** (c. 7.1 days).

Saturn is in **Aquarius** at the start of the year, it moves into **Pisces** in April, then back into Aquarius in September. On 13 July Saturn will enter retrograde motion in Pisces. It will reach opposition on 21 September and it will shine brightly with a magnitude of 0.6. Saturn will end its retrograde movement on 28 November, gradually getting fainter and dipping to mag. 1.0 in December. On 23 March Saturn's rings will appear edge-on as viewed from Earth. The planet will be too close to the Sun at this time, it will reappear in the early morning sky from late June onwards, setting earlier towards the end of the year.

The path of Saturn in 2025. Saturn comes to opposition on 21 September. Background stars are shown down to magnitude 6.5.

Uranus and Neptune

Uranus ends its retrograde motion on 30 January (it has been moving westward since September 2024) while in **Aries**. It will have a mag. of 5.7. Uranus moves into **Taurus** in March. It will re-enter retrograde motion on 6 September 2025, moving westward across the sky until 4 February 2026. It will reach opposition on 21 November in Taurus, mag. 5.6.

Neptune moved into **Pisces** in May 2022; it will stay in Pisces until August 2029. It enters retrograde motion on 4 July 2025 at magnitude 7.8. It will reach opposition on 23 September (mag. 7.7). It reverts to direct motion on 10 December and will continue moving eastwards until 7 July 2026.

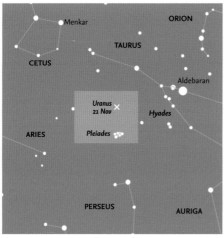

Uranus is in Taurus at the time of its opposition and not far from the Pleiades, on 21 November. The boxed area is shown in more detail to the right.

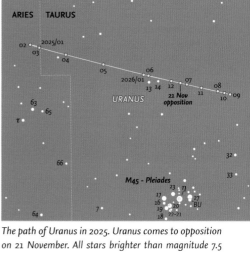

The path of Uranus in 2025. Uranus comes to opposition on 21 November. All stars brighter than magnitude 7.5 are shown.

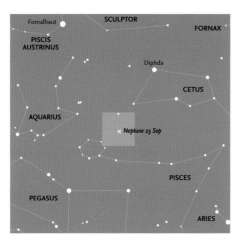

In 2025, Neptune is to be found in the southern part of Pisces. The planet comes to opposition on 23 September.

The path of Neptune in 2025. It is in the constellation of Pisces for the whole year. All stars down to magnitude 8.5 are shown.

Minor Planets in 2025

Several minor planets rise above magnitude 9.0 in 2025, and four come to opposition at, or above, mag. 9.0. They are closest to Earth at this point and reach their highest point in the sky around midnight – this is the best time to see them. On 2 May the asteroid *(4) Vesta* will lie in **Libra**, reaching a peak magnitude of 5.6 at opposition. On 11 August *(89) Julia* will reach magnitude 8.5 in **Aquarius** and *(6) Hebe* will reach opposition later in the month on 26 August, visible at magnitude 7.6.

On 2 October the dwarf planet *(1) Ceres* will rise in magnitude to 7.6, at a distance of 1.97 AU (Astronomical Units, equivalent to the average distance from the Earth to the Sun), though its visibility will be hindered by a waxing gibbous Moon. Ceres is the only dwarf planet that lies within the orbit of Neptune and it completes its orbit in 4.6 years. The Earth could accommodate 2,500 of Ceres.

The three small maps at the top of the next page show the areas covered by the larger maps in a lighter blue. A white cross marks the position of the minor planet at the day of its opposition. South is up on all maps.

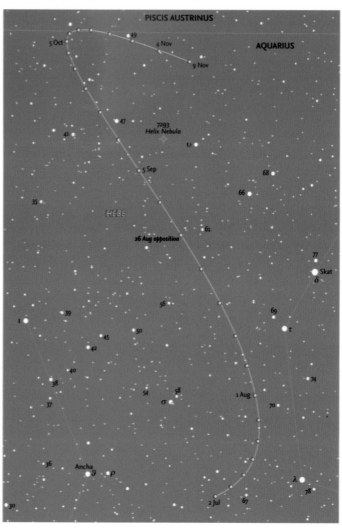

The path of the minor planet (6) Hebe around its opposition on 26 August (mag. 7.6). Background stars are shown down to magnitude 9.5.

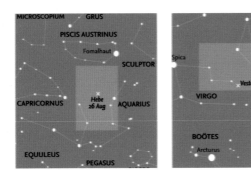

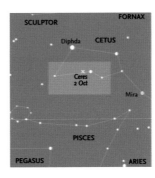

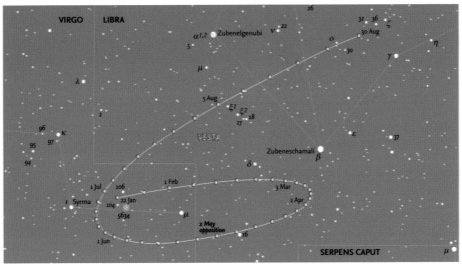

The path of the minor planet (4) Vesta around its opposition on 2 May (mag. 5.6). Background stars are shown down to magnitude 8.5.

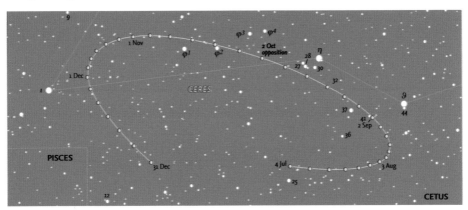

The path of the minor planet (1) Ceres around its opposition on 2 October (mag. 7.6). Background stars are shown down to magnitude 9.5.

Comets in 2025

Although comets may occasionally become very striking objects in the sky, as was the case with Comet C/2020 F3 NEOWISE in 2020, their occurrence and particularly the existence or length of any tail and their overall magnitude are notoriously difficult to predict. Naturally, it is only possible to predict the return of periodic comets (whose names have the prefix 'P'). Many comets appear unexpectedly (these have names with the prefix 'C'). Bright, readily visible comets such as C/1995 Y1 Hyakutake, C/1995 O1 Hale-Bopp, C/2006 P1 McNaught or C/2020 F3 NEOWISE are rare. Most periodic comets are faint and only a very small number ever become bright enough to be easily visible with the naked eye or with binoculars.

Comet 24P/Schaumasse will be visible after midnight in December 2025, moving through the constellations of Leo, Coma Berenices and then passing into Virgo in early January 2026. It will make its closest approach on 4 January 2026, passing within a distance equivalent to 60% of the Earth-Sun distance. It will brighten towards the end of the year, reaching a magnitude of between 8 and 9.

The Comet C/2020 F3 NEOWISE photographed by Nick James on 17 July 2020, showing the light-coloured dust tail, together with the blue ion tail.

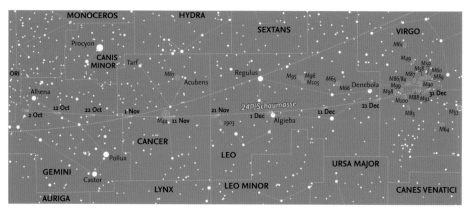

The path of Comet 24P/Schaumasse from 2 October to 31 December 2025. It rises above magnitude 9 in December. Background stars are shown down to magnitude 6.0. South is up.

Comet 12P/Pons-Brooks, showing the coma (halo) of gas surrounding the nucleus and a tail.

Introduction to the Month-by-Month Guide

The monthly charts

The pages devoted to each month contain a pair of charts showing the appearance of the night sky, looking north and looking south. The charts (as with all the charts in this book) are drawn for the latitude of 35°S, so observers farther south will see slightly more of the sky on the southern horizon, and slightly less on the northern. These areas are, of course, those most likely to be affected by poor observing conditions caused by haze, mist or smoke. In addition, stars close to the horizon are always dimmed by atmospheric absorption, so sometimes the faintest stars marked on the charts may not be visible.

The three times shown for each chart require a little explanation. The charts are drawn to show the appearance at 23:00 UT for the 1st of each month. The same appearance will apply an hour earlier (22:00 UT) on the 15th, and yet another hour earlier (21:00 UT) at the end of the month (shown as the 1st of the following month). These times, where UT stands for the Universal Time used by astronomers around the world, are identical to GMT. Daylight Saving Time (DST) is not used in South Africa. In New Zealand, it applies from the last Sunday of September to the first Sunday of April. In those Australian states with DST, it runs from the first Sunday in October to the first Sunday in April. The appropriate times are shown on the monthly charts. Times of specific events are shown on the 24-hour clock of Universal Time (UT), used by astronomers worldwide, and corrections for the local time zone

The charts may be used for earlier or later times during the night. To observe two hours earlier, use the charts for the preceding month; for two hours later, the charts for the next month.

Meteors

Details of specific meteor showers are given in the months when they come to maximum,

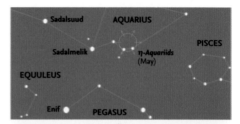

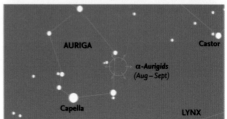

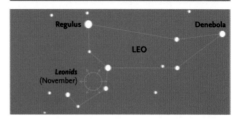

regardless of whether they begin or end in other months. Note that not all the respective radiants are marked on the charts for that particular month, because the radiants may be below the horizon, or lie in constellations that are not readily visible during the month of maximum. For this reason, special charts for the Eta (η) Aquariids (May), the Alpha (α) Aurigids (August and September) and the Leonids (November) are given here. As just explained, however, meteors from such showers may still be seen, because the most effective region for seeing meteors is some 40–45° away from the radiant, and that area of sky may well be above the horizon. A table of the best meteor showers visible during the year is also given here. The rates given are based on the properties of the meteor streams, and are those that an experienced observer might see under ideal conditions. Generally, the observed rates will be far lower.

Shower	Dates of activity 2025	Date of maximum 2025	Possible hourly rate
α-Centaurids	28 January to 21 February	8 February	6
γ-Normids	25 February to 28 March	14 March	6
π-Puppids	15 April to 28 April	23 April	var.
η-Aquariids	19 April to 28 May	5 May	40
α-Capricornids	3 July to 15 August	30 July	5
Southern δ-Aquariids	12 July to 23 August	30 July	25
Piscis Austrinids	15 July to 10 August	28 July	5
Perseids	17 July to 24 August	12 August	150
α-Aurigids	28 August to 5 September	31 August	6
Southern Taurids	10 September to 20 November	10 October	5
Orionids	2 October to 7 November	22 October	15
Northern Taurids	20 October to 10 December	12 November	5
Leonids	6 November to 30 November	17 November	15
Phoenicids	28 November to 9 December	2 December	var.
Puppid-Velids	1 December to 15 December	7 December	10
Geminids	4 December to 20 December	14 December	120

Meteors that are brighter than magnitude -4 (approximately the maximum magnitude reached by Venus) are known as *fireballs* or *bolides*. Examples are shown on pages 32, 35 and 77. Fireballs sometimes cause sonic booms that may be heard some time after the meteor is seen.

The photographs

As an aid to identification – especially as some people find it difficult to relate charts to the actual stars they see in the sky – one or more photographs of constellations visible in certain specific months are included. It should be noted, however, that because of the limitations of the photographic and printing processes, and the differences between the sensitivity of different individuals to faint starlight (especially in their ability to detect different colors), and the degree to which they have become adapted to the dark, the apparent brightness of stars in the photographs will not necessarily precisely match that seen by any one observer.

The Moon calendar

The Moon calendar is largely self-explanatory. It shows the phase of the Moon for every day of the month, with the exact times (in Universal Time) of New Moon, First Quarter, Full Moon and Last Quarter. Because the times are calculated from the Moon's actual orbital parameters, some of the times shown

will, naturally, fall during daylight, but any difference is too small to affect the appearance of the Moon on that date. Also shown is the *age* of the Moon (the day in the *lunation*), beginning at New Moon, which may be used to determine the best time for observation of specific lunar features.

The Moon

The section on the Moon includes details of any lunar or solar eclipses that may occur during the month (visible from anywhere on Earth). Similar information is given about any important occultations. Mainly, however, this section summarizes when the Moon passes close to planets or the five prominent stars close to the ecliptic. The dates when the Moon is closest to the Earth (at *perigee*) and farthest from it (at *apogee*) are shown in the monthly calendars, and only mentioned here when they are particularly significant, such as the nearest and farthest points during the year.

The planets and minor planets

Brief details are given of the location, movement and brightness of the planets from Mercury to Saturn throughout the month. None of the planets can, of course, be seen when they are close to the Sun, so such periods are generally noted. All of the planets may sometimes lie on the opposite side of the Sun to the Earth (at superior conjunction),

A fireball (with flares approximately as bright as the Full Moon), photographed against a weak auroral display from Tarbat Ness by Denis Buczynski in Scotland on 22 January 2017.

but in the case of the inferior planets, Mercury and Venus, they may also pass between the Earth and the Sun (at inferior conjunction) and are normally invisible for a period of time, the length of which varies from conjunction to conjunction. Those two planets are normally easiest to see around either eastern or western elongation, in the evening or morning sky, respectively. Not every elongation is favorable, so although every elongation is listed, only those where observing conditions are favorable are shown in the individual diagrams of events.

The dates at which the superior planets reverse their motion (from direct motion to *retrograde*, and retrograde to direct) and of opposition (when a planet generally reaches its maximum brightness) are given. Some planets, especially distant Saturn, may spend most or all of the year in a single constellation. Jupiter and Saturn are normally easiest to see around opposition, which occurs every year. Mars, by contrast, moves relatively rapidly against the background stars and in some years never comes to opposition.

Uranus is normally magnitude 5.7–5.9, and thus at the limit of naked-eye visibility under exceptionally dark skies, but bright enough to be readily visible in binoculars. Because its orbital period is so long (over 84 years), Uranus moves only slowly along the ecliptic, and often remains within a single constellation for a whole year. The chart on page 25 shows its position during 2025.

Similar considerations apply to Neptune, although it is always fainter (generally magnitude 7.8–8.0), it is still visible in most binoculars. It takes about 164.8 years to complete one orbit of the Sun. As with Uranus, it frequently spends a complete year in one constellation. Its chart is also on page 25.

In any year, few minor planets ever become bright enough to be detectable in binoculars. Just one, (4) Vesta, on rare occasions brightens sufficiently for it to be visible to the naked eye. Our limit for visibility is magnitude 9.0 and details and charts are given for those objects that exceed that magnitude during the year, normally around opposition. To assist in recognition of a planet or minor planet as it moves against the background stars, the latter are shown to a fainter magnitude than the object at opposition. Minor-planet charts for 2025 are on pages 26 and 27.

The ecliptic charts

Although the ecliptic charts are primarily designed to show the positions and motions of the major planets, they also show the motion of the Sun during the month. The light-tinted area shows the area of the sky that is invisible during daylight, but the darker area gives an indication of which constellations are likely to be visible at some time of the night. The closer a planet is to the border between dark and light, the more difficult it will be to see in the twilight.

The monthly calendar

For each month, a calendar shows details of significant events, including when planets are close to one another in the sky, close to the Moon, or close to any one of five bright stars that are spaced along the ecliptic. The times shown are given in Universal Time (UT), always used by astronomers throughout the year, and which is identical to Greenwich Mean Time (GMT). So during the summer months, they do not show Summer Time, which will need to be taken for the observing location.

The diagrams of interesting events

Each month, a number of diagrams show the appearance of the sky when certain events take place. However, the exact positions of celestial objects and their separations greatly depend on the observer's position on Earth. When the Moon is one of the objects involved, because it is relatively close to Earth, there may be very significant changes from one location to another. Close approaches between planets or between a planet and a star are less affected by changes of location, which may thus be ignored.

The diagrams showing the appearance of the sky are drawn at latitude 35°S and longitude 150°E (approximately that of Sydney, Australia), so will be approximately correct for much of Australia. However, for an observer farther north (say, in Brisbane or Darwin), a planet or star listed as being north of the Moon will appear even farther north, whereas one south of the Moon will appear closer to it – or may even be hidden (occulted) by it. For an observer at a latitude greater than 35°S (such as in New Zealand), there will be corresponding changes in the opposite direction: for a star or planet south of the Moon, the separation will increase, and for one north of the Moon, the separation will decrease. This is particularly important when occultations occur, which may be visible from one location, but not another. There are three occultations of Mars visible in 2025, but only one is visible from the southern hemisphere (see the monthly calendars for more details). Ideally, details should be calculated for each individual observer, but this is obviously impractical. In fact, positions and separations are calculated for a theoretical observer located at the centre of the Earth.

So the details given regarding the positions of the various bodies should be used as a guide to their location. A similar situation arises with the times that are shown. These are calculated according to certain technical criteria, which need not concern us here. However, they do not necessarily indicate the exact time when two bodies are closest together. Similarly, dates and times are given, even if they fall in daylight, when the objects are likely to be completely invisible. However, such times do give an indication that the objects concerned will be in the same general area of the sky during both the preceding and the following nights.

Data used in this Guide

The data given in this Guide, such as timings and distances between objects, have been computed with a program developed by the US Naval Observatory in Washington DC (MICA, the Multiyear Interactive Computer Almanac). Additional data was provided by the following sources: the Astronomical Applications Department of the US Naval Observatory; the Sky Events Calendar by Fred Espenak and Sumit Dutta (NASA's GSFC); the NASA Jet Propulsion Laboratory (JPL) Horizons System; the International Astronomical Union's (IAU) Minor Planet Center.

Key to the symbols used on the monthy star maps.

STAR MAGNITUDES								PLANETS	
-1	0	1	2	3	4	Fainter	*Milky Way*		MERCURY
									VENUS
OBJECTS									MARS
Open Star Cluster			*Planetary Nebula*		*Meteor Radiant*				JUPITER
Globular Star Cluster			*Bright Nebula*		*Galaxy*				SATURN

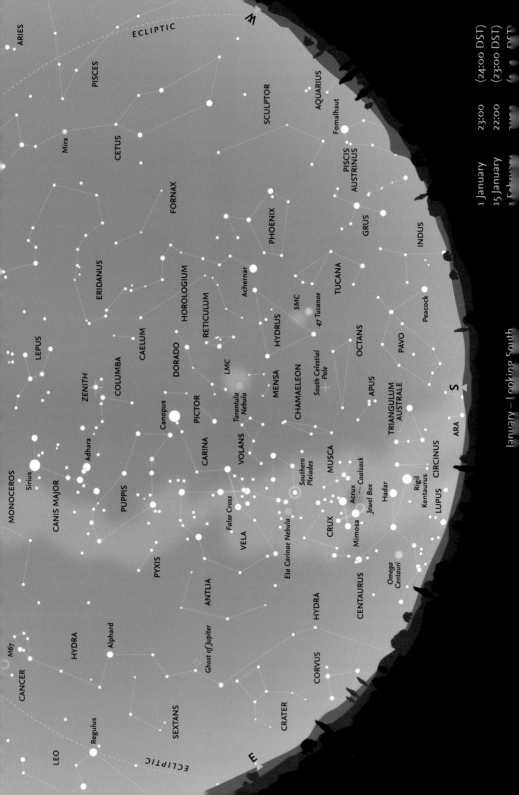

January – Looking South

January – Looking South

Crux is now quite prominent as it rises in the east, and even **Rigil Kentaurus** and **Hadar** (α and β Centauri), although low, are becoming easier to see. The whole of **Carina** is visible, with both the **Eta Carinae Nebula** and the **Southern Pleiades** between Crux and the **False Cross** of stars from the constellations of Carina and **Vela**. **Canopus** (α Carinae) is high in the south, roughly three-quarters of the way from the horizon to the zenith. Also in the south, but slightly lower, is the **Large Magellanic Cloud** (LMC). **Achernar** (α Eridani) is prominent between the constellations of **Hydrus** and **Phoenix**. The **Small Magellanic Cloud** (SMC) and **47 Tucanae** are well-placed for observation alongside Hydrus. **Fomalhaut** (α Piscis Austrini) may be glimpsed low on the horizon towards the west as may **Peacock** (α Pavonis) farther towards the south. The whole of **Pavo** and **Grus** is visible, together with most of **Indus.**

Meteors

The year does not start well for southern meteor observers, with the only shower being the **α-Centaurids** (active late January to February), which have low maximum rates, as do the **γ-Normids** (February to March).

Part of the southern Milky Way. The two bright stars near the bottom of the image are α and β Centauri. The constellation of Crux is near the centre. The Eta Carinae Nebula and the Southern Pleiades are near the top-right corner.

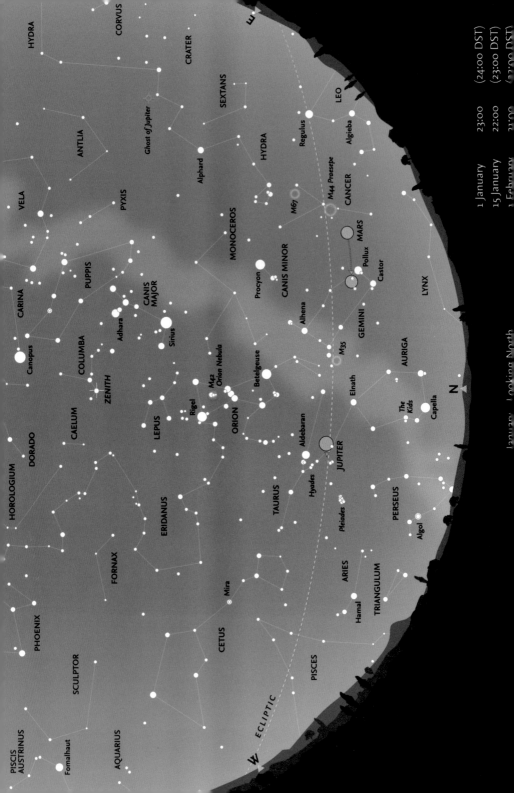

January. Looking North

1 January	23:00	(24:00 DST)
15 January	22:00	(23:00 DST)
1 February	21:00	(22:00 DST)

January – Looking North

The northern sky is dominated by **Orion**, prominent during the summer months and visible at some time during the night. It is highly distinctive, with a line of three stars that form the Orion's Belt. To most observers, the bright star **Betelgeuse** (α Orionis) shows a reddish tinge, in contrast to the brilliant bluish-white **Rigel** (β Orionis). The three stars of the belt lie directly south of the celestial equator. A vertical line of three 'stars' forms the 'Sword' that hangs south of the belt. With good viewing, the central 'star' appears as a hazy spot, even to the naked eye, and is actually the **Orion Nebula**. Binoculars reveal the four stars of the Trapezium, which illuminate the nebula.

The Belt points down to the northwest towards **Taurus** (the Bull) and orange-tinted **Aldebaran** (α Tauri). Close to Aldebaran, there is a conspicuous 'V' of stars, called the **Hyades** cluster. (Despite appearances, Aldebaran is not part of the cluster.) Farther along, the same line from Orion passes above a bright cluster of stars, the **Pleiades**, or Seven Sisters. Even the smallest pair of binoculars reveals this as a beautiful group of bluish-white stars. The two most conspicuous of the other stars in Taurus lie directly below Orion, and form an elongated triangle with Aldebaran. The northernmost, **Elnath** (β Tauri), was once considered to be part of the constellation of **Auriga.**

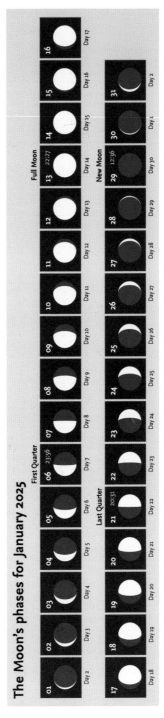

The constellation of Orion dominates the northern sky during this period of the year and is a useful starting point for recognizing other constellations in the area (see page 36). Here, orange Betelgeuse, blue-white Rigel and the pinkish Orion Nebula are prominent (south is up).

The Moon's phases for January 2025

						First Quarter						Full Moon			
01	02	03	04	05	06 23:56	07	08	09	10	11	12	13 22:27	14	15	16
Day 1	Day 2	Day 3	Day 4	Day 5	Day 6	Day 7	Day 8	Day 9	Day 10	Day 11	Day 12	Day 13	Day 14	Day 15	Day 16

				Last Quarter									New Moon		
17	18	19	20	21 20:31	22	23	24	25	26	27	28	29 12:36	30	31	
Day 17	Day 18	Day 19	Day 20	Day 21	Day 22	Day 23	Day 24	Day 25	Day 26	Day 27	Day 28	Day 29	Day 30	Day 1	Day 2

January – Moon and Planets

The Moon

On 3 January, a few days before First Quarter, the Moon will be only 1.4° from **Venus**, visible in the early evening. Venus will be mag. -4.4. A day later the Moon will occult **Saturn**, visible from Europe, Africa and western Russia. On 6 January the Moon will be at First Quarter, setting around midnight. On 10 January the waxing gibbous Moon will pass within 5.4°N of **Jupiter** (mag. -2.7), visible all night until just after 04:00 GMT. On 13 January the Full Moon will be 2.1°S of **Pollux**. One day later, **Mars** and the Moon will be together in the morning sky, less than half a degree apart, with Mars at mag. -1.4. Two days later the Moon will be 2.2°N of **Regulus**. The Moon will be a tenth of a degree from **Spica** on 21 January, and on the dawn of the 24th it will appear only 0.3°S of **Antares**. On 29 January the Moon

passes between the Sun and the Earth, visible as a New Moon. As the month ends, the Moon will appear as a waxing crescent, setting shortly after the Sun.

The planets

Venus comes to greatest eastern elongation on 10 January (47.2°E), when it will be visible at mag. -4.4. **Mars** will reach perigee on 12 January, passing within a distance of 0.64 AU of the Earth. The best time to observe Mars will be close to midnight on 16 January when it reaches opposition. **Jupiter** will lie in **Taurus**, on 31 January it will be 5.2°N of **Aldebaran**. On 20 January **Saturn** and **Venus** will be 2.2° apart, with Venus the brighter of the two at mag. -4.5. Saturn will be in **Aquarius** and visible until just after 20:00 GMT. On 30 January **Uranus** ends its westward retrograde motion while in **Aries**. **Neptune** will lie in **Pisces**, setting earlier in the evening towards the end of the month.

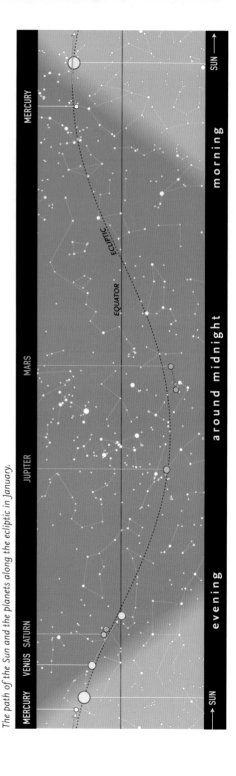

The path of the Sun and the planets along the ecliptic in January.

Calendar for January

03	15:24	Venus (mag. –4.4) 1.4°N of Moon
04	14:00	Earth at perihelion: 0.9833 AU
04	17:18	Saturn (mag. 1.1) 0.7°S of Moon
06	23:56	FIRST QUARTER MOON
07	23:34	Moon at perigee: 370,173 km
10	01:01	Pleiades 0.3°S of Moon
10	04:00	Venus at greatest elongation: 47.2°E
10	23:13	Jupiter (mag. –2.7) 5.4°S of Moon
13	21:45	Pollux 2.1°N of Moon
13	22:27	FULL MOON
14	03:42	Mars (mag. –1.4) 0.2°S of Moon. An occultation will be visible from some parts of Northern America and Africa.
16	01:00	Mars at opposition (mag. –1.4)
16	14:57	Regulus 2.2°S of Moon
18	16:00	Venus (mag.–4.6) 2.2°N of Saturn (mag. 1.1)
19	14:00	Mercury at aphelion
21	03:53	Spica 0.1°N of Moon
21	04:55	Moon at apogee: 404,299 km
21	20:31	LAST QUARTER MOON
23	17:07	Mars (mag. –1.3) 2.3°S of Pollux
24	23:34	Antares 0.3°N of Moon
29	12:36	NEW MOON

Evening 21:00 (DST)

3–4 January • The crescent Moon moves from bright Venus (mag. –4.4) to the much fainter Saturn (mag. 1.1). Enif (ε Peg) is very close to the horizon.

9–11 January • The Moon passes just south of the Pleiades. It moves along Aldebaran and Jupiter (mag. –2.7) and is close to Elnath (β Tau) on 11 January.

After midnight 1:00 (DST)

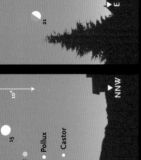

Morning 4:00 (DST)

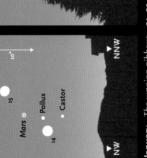

14–15 January • The waning gibbous Moon passes Castor, Pollux and Mars (mag. –1.4).

21–22 January • Shortly after Last Quarter, the Moon moves past Spica, low in the east.

Early morning 3:00 (DST)

25 January • The Moon is in the constellation of Scorpius, close to bright red Antares.

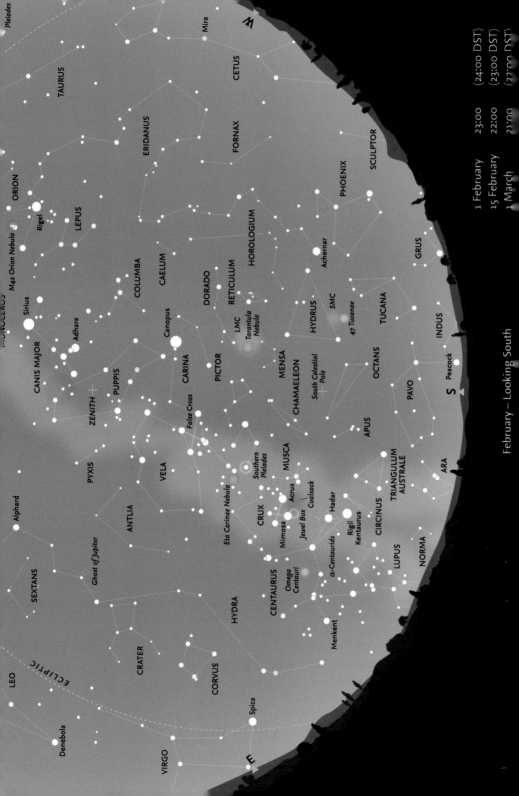

February – Looking South.

February – Looking South

The whole of **Centaurus** is now clear of the horizon in the southeast, where, below it, the constellation of **Lupus** is beginning to be visible. **Triangulum Australe** is now higher in the sky and easier to see. Both **Crux** and the **False Cross** are also higher and more visible. The **Large Magellanic Cloud** (LMC) and brilliant **Canopus** are now almost due south, with the constellation of **Puppis** at the zenith. Both **Hydrus** and the **Small Magellanic Cloud** (SMC) are lower in the sky, as is **Phoenix**, which is much closer to the horizon in the west. Next to it, the constellation of **Tucana** is also lower although the globular cluster **47 Tucanae** remains readily visible as does **Achernar** (α Eridani). The constellation of **Grus** and bright **Fomalhaut** (α Piscis Austrini) have disappeared below the horizon. **Peacock** (α Pavonis) is skimming the horizon in the south and is not easily seen at any time in the night.

Meteors

The **Centaurid** shower (which actually consists of two separate streams: the **α-** and **β-Centaurids**), whose radiants lie near α and β Centauri (Rigil Kentaurus and Hadar), respectively – continues in February. The shower reaches a low maximum of around 6 meteors per hour on 8 February, when the Moon is a bright waxing gibbous, setting a few hours after midnight. Another weak shower, the **γ-Normids**, begins to be active on 25 February, but the meteors are difficult to differentiate from sporadics. It reaches its weak (but sharp) maximum on 14 March.

Parts of Carina and Vela. The 'False Cross' is indicated with blue lines. The red blurry spot near the bottom-left corner is the Eta (η) Carinae Nebula. To the right, and a little lower, is an open cluster that surrounds Theta (θ) Carinae. This cluster (IC 2602) is also known as the Southern Pleiades.

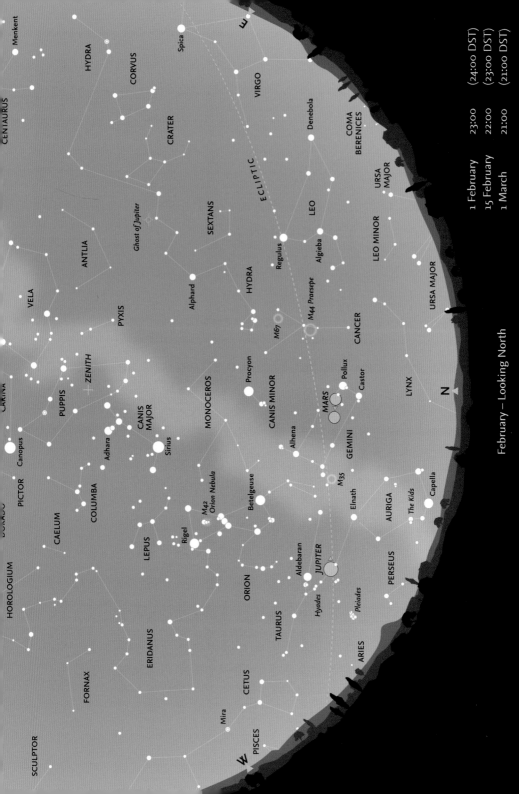

February – Looking North

February – Looking North

The constellation of **Gemini**, with the pair of stars **Castor** and **Pollux** (α and β Geminorum) is now due north. Pollux is higher in the sky (farther towards the zenith), and above it is **Procyon** (α Canis Minoris), halfway to the zenith. Still higher is **Sirius**, the brightest star in the sky (at mag. -1.4) and in the constellation of **Canis Major**. The faint constellation of **Cancer**, with its most noticeable feature, the cluster **Praesepe**, lies to the east of Gemini. Above it, and directly east of Procyon, is the distinctive asterism forming the head of **Hydra**, the whole of which constellation is now visible stretching across the sky towards the east. The constellation of **Taurus**, with orange **Aldebaran** (α Tauri) is still clearly seen in the west. By contrast, **Auriga** is much lower towards the horizon and brilliant **Capella** (α Aurigae) is extremely low and visible only early in the night. The faintest stretch of the Milky Way runs from Auriga in the northwest up towards the zenith, passing through Gemini and the indistinct constellation of **Monoceros**. In the northeast, the zodiacal constellation of **Leo** and bright **Regulus** (α Leonis) are clearly seen.

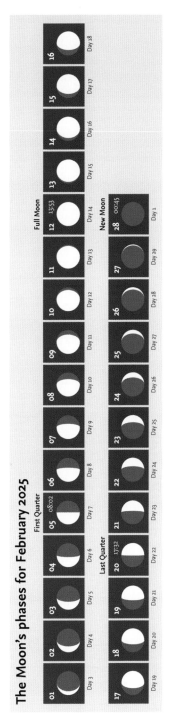

The constellation of Cancer. Almost at the centre of the image is the open star cluster M44 (Praesepe), sometimes called the 'Beehive'. The faint constellation of Cancer is surrounded by several bright stars: Castor and Pollux, near the bottom, Procyon near the top and Regulus with Algieba close to the right edge of the image (south up).

The Moon's phases for February 2025

First Quarter

01	02	03	04	05 08:02	06	07	08
Day 3	Day 4	Day 5	Day 6	Day 7	Day 8	Day 9	Day 10

Last Quarter

17	18	19	20 17:32	21	22	23	24
Day 19	Day 20	Day 21	Day 22	Day 23	Day 24	Day 25	Day 26

Full Moon

09	10	11	12 13:53	13	14	15	16
Day 11	Day 12	Day 13	Day 14	Day 15	Day 16	Day 17	Day 18

New Moon

25	26	27	28 00:45				
Day 27	Day 28	Day 29	Day 1				

February – Moon and Planets

The Moon

On 1 February the thin waxing crescent Moon will be 1° from **Saturn** (mag. 1.0) and just over 2° from **Venus** (mag. -4.6), visible until 3 hours after sunset. First Quarter Moon will be on 5 February and two days later it will be seen 5.5°N of **Jupiter** (mag. -2.5) until a few hours after midnight. On the 9th the Moon will occult **Mars**, an event that will be visible from the northwest of Great Britain, Russia, China and eastern Canada. From London they will be less than a degree apart. The following day the Moon will be 2.1° S of **Pollux** and two days later the Full Moon will appear 2.2°N of **Regulus**. On 17 February it will be **Spica's** turn, less than half a degree from the waning gibbous Moon. The Last Quarter Moon is on 20 February and the following day the waning crescent Moon will be half a degree from orange-red **Antares**. The month ends with a New Moon on the 28th.

The planets

Mercury (mag. -1.0 to -1.6) moves from **Aquarius** to **Pisces** on 26 February, setting before the Sun until 9 February, after which it will set progressively later than the Sun. **Venus** will reach its highest point in the sky on 7 February, reaching greatest brightness on 16 February at mag. -4.2 and setting several hours after the Sun. **Mars** will be visible with its orange hue all night, shining brighter at the beginning of the month (mag. -1.1 to -0.3). On 4 February **Jupiter** (mag. -2.5) will reach the end of its retrograde motion (in **Taurus**), ending its westward movement through the constellations and returning to its usual eastward motion. **Saturn** (mag. 1.1) in **Aquarius** will be low in the western sky shortly after sunset. **Uranus** (mag. 5.7 to 5.8) will be high in **Aries** a few hours before midnight. Neptune (mag. 7.8) is in **Pisces.**

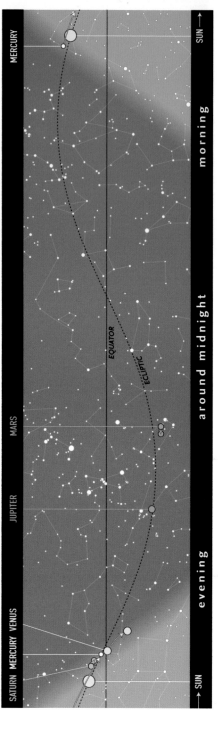

The path of the Sun and the planets along the ecliptic in February.

Calendar for February

01	04:46	Saturn (mag. 1.0) 1.1°S of Moon
01	20:27	Venus (mag. -4.6 2.3°N of Moon
02	02:43	Moon at perigee: 367,457 km
05	08:02	FIRST QUARTER MOON
06	06:43	Pleiades 0.5°S of Moon
07	03:35	Jupiter (mag. -2.5) 5.5°S of Moon
08		α-Centaurids Meteor Shower maximum
09	19:36	Mars (mag. -0.9) 0.8°S of Moon
10	05:19	Pollux 2.1°N of Moon
12	13:53	FULL MOON
12	23:21	Regulus 2.2°S of Moon
17	12:01	Spica 0.3°N of Moon
18	01:11	Moon at apogee: 404,882 km
19	18:00	Venus at perihelion
20	17:32	LAST QUARTER MOON
21	08:21	Antares 0.4°N of Moon
28	00:45	NEW MOON

Evening 20:45 (DST)

Evening 23:00 (DST)

1–2 February • Shortly after sunset, the crescent Moon passes Saturn (mag. 1.1) and bright Venus (mag. -4.7), in the western sky.

6–7 February • Late in the evening, the Moon is in the northwestern sky and passes the Pleiades, Aldebaran and Jupiter (mag. -2.5).

After midnight 2:00 (DST)

Early morning 5:00 (DST)

After midnight 2:00 (DST)

10 February • The Moon is in the company of Mars (mag. -0.8) and the Twin Stars, Castor and Pollux.

13–14 February • Low in the northwest, the Full Moon passes between Regulus and Algieba (γ Leo).

21–22 February • The Moon passes Acrab (β Sco), Dschubba (δ Sco), and the red star Antares.

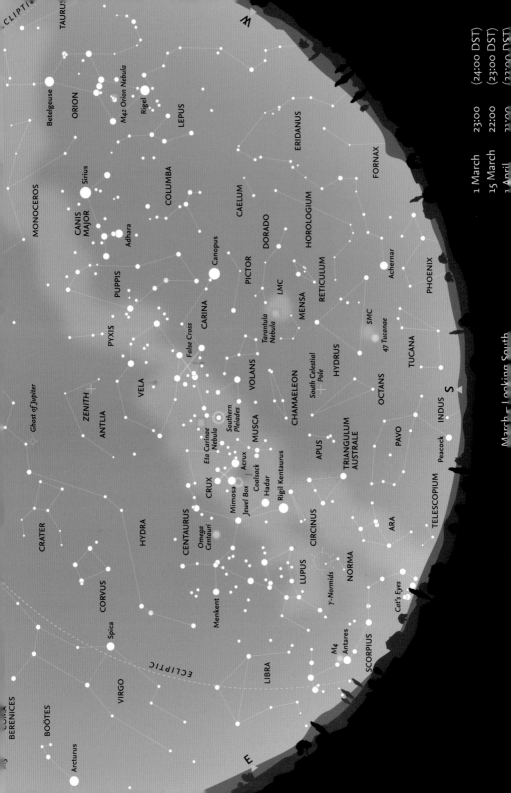

March – Looking South

1 March	23:00	(24:00 DST)
15 March	22:00	(23:00 DST)
1 April	21:00	(22:00 DST)

March – Looking South

The constellation of Vela may be found in the lower part of the image, as well as the False Cross, consisting of two stars that belong to Vela, while the two stars near the bottom are part of Carina. The constellation of Pyxis is near the top (north is up).

The Sun crosses the celestial equator on Thursday 20 March, at the equinox, when day and night are of almost equal length, and the southern season of autumn is considered to begin. (The hours of daylight and darkness change most rapidly around the equinoxes in March and September.)

Scorpius is beginning to become visible in the eastern sky, including the 'Cat's Eyes', the pair of stars **Shaula** and **Lesath** (λ and υ Scorpii, respectively) at the end of the 'sting'. Above Scorpius, the whole of the constellation of **Lupus** is now easy to see. The magnificent globular cluster of **Omega Centauri** is now readily visible, northeast of **Crux**. The **Coalsack** and the denser region of the Milky Way in **Carina**, together with the **Eta Carinae Nebula** and the **Southern Pleiades** are well placed for observation. The **False Cross** on the **Carina/Vela** border is now high in the sky, between the South Celestial Pole and the zenith. Brilliant **Canopus** (α Carinae) is only slightly lower towards the west, above the **Large Magellanic Cloud** (LMC) and the striking **Tarantula Nebula**. **Achernar** (α Eridani), the **Small Magellanic Cloud** (SMC) and **47 Tucanae** are considerably lower, but still clear of the horizon. **Peacock** (α Pavonis) remains low, skimming the horizon, just east of south. **Orion** is now visibly getting lower in the west, and is being followed by **Sirius** and **Canis Major**.

Meteors

The only significant meteor shower in March is the **γ-Normids**, which have a low rate, and are thus difficult to differentiate from sporadics. However, they exhibit a very sharp peak a day or so on either side of maximum on 14 March, at Full Moon, so conditions are unfavourable. The faint constellation of **Norma** rises early in the night, but most meteors are likely to be seen (away from the radiant) in the hours after midnight.

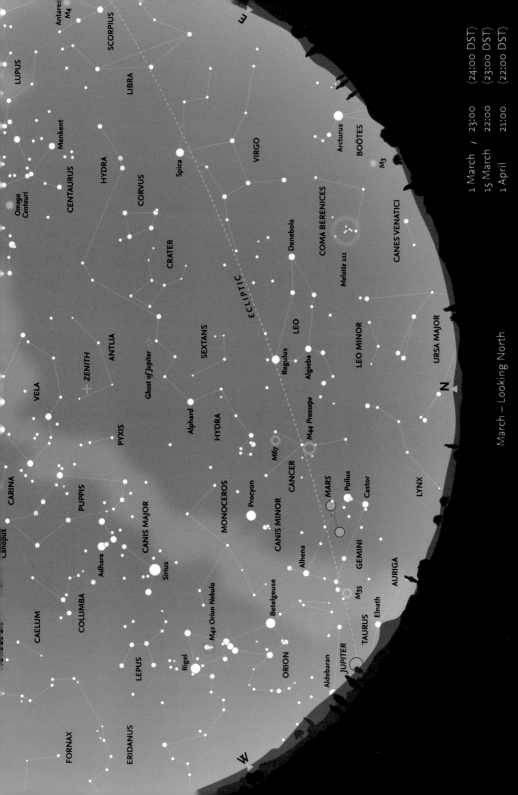

March – Looking North

1 March / 23:00 (24:00 DST)
15 March 22:00 (23:00 DST)
1 April 21:00 (22:00 DST)

March – Looking North

Almost due north is the constellation of Leo, with the 'backward question mark' (or 'Sickle') of bright stars forming the head of the mythological lion. **Regulus** (α Leonis) – the 'dot' of the question mark or the handle of the sickle and the brightest star in Leo – lies very close to the ecliptic and is one of the few first-magnitude stars that may be occulted by the Moon, although none occur in 2025. The Moon often passes between Regulus and **Algieba** (γ Leonis). To the west lies the faint constellation of **Cancer**, with the open cluster M44, or **Praesepe**. The constellations of **Gemini** and **Orion** are now getting low in the west, but above them, both **Procyon** in **Canis Minor** and the constellation of **Canis Major** remain clear to see. The whole of **Hydra** (the largest constellation) now sprawls right across the southern and eastern skies, and the three small constellations of **Sextans**, **Crater** and **Corvus** are readily visible. The unremarkable constellation of **Antlia** is at the zenith. In the east, **Arcturus** (α Boötis) – the brightest star in the northern hemisphere of the sky – is becoming visible, and climbs higher during the night.

The distinctive constellation of Leo, with Regulus and 'the Sickle' on the west. Algieba (γ Leonis), north of Regulus, appears double, but is a multiple system of four stars. (south is up).

The Moon's phases for March 2025

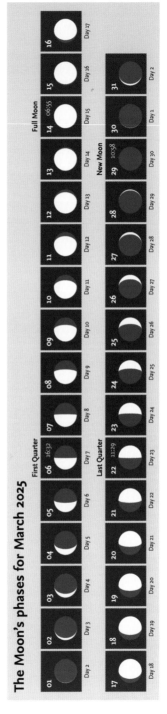

March – Moon and Planets

The Moon

On 1 March the thin waxing crescent Moon will lie 0.4°S of Mercury, and an occultation will be visible from some parts of Australia. The Moon will also be just over 6° from **Venus** (mag. -4.6) in the early evening sky. On 6 March the First Quarter Moon will pass 5.6°N of **Jupiter** (mag. -2.3). On 9 March the waxing gibbous Moon will be visible less than 2° apart from **Mars** (mag. -0.1) and 2.0°S of **Pollux**. Three days later the bright waxing gibbous Moon will be 2.2°S of **Regulus**. A total lunar eclipse on 14 March will be visible from some western parts of Europe, the Americas, north-eastern Russia and Africa – a partial eclipse will be visible from some regions in eastern Australia. On 16 March the Moon and **Spica** will be less than half a degree apart and an occultation of Spica will be visible from

Australia and parts of Africa. On 20 March the waning Moon is only 0.5°S of **Antares**, visible at dawn. Last Quarter Moon is on 22 March. The Moon and Saturn (mag. 1.1) will be less than 2° apart on 28 March, setting with the Sun. The New Moon occurs on 29 March.

The Planets

Mercury will reach its highest point in the sky on 8 March (mag. -0.4). One day later **Venus** (mag. -4.4) will pass 6.3°N of Mercury (mag. -0.2), visible at dusk. **Mars** (mag. -0.2 to 0.4) is still easily visible throughout the first half of the night. **Jupiter** (mag. -2.3 to -2.1) is also best seen before midnight, placed in **Taurus**. **Saturn** (mag. 1.1 to 1.2), lying in **Aquarius**, sets with the Sun. **Uranus** (mag. 5.8) lies in **Aries** and is above the horizon a few hours before midnight. **Neptune** (mag. 7.8) is in **Pisces**.

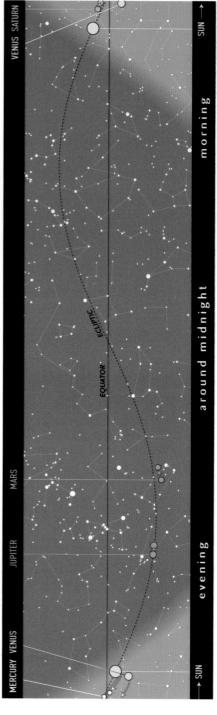

The path of the Sun and the planets along the ecliptic in March.

Calendar for March

01	04:03	Mercury (mag. −0.1) 0.4°N of Moon. An occultation will be visible from some parts of Australia.
01	21:18	Moon at perigee: 361,967 km
04	14:00	Mercury at perihelion
05	12:32	Pleiades 0.6°S of Moon
06	11:31	Jupiter (mag. −2.3) 5.6°S of Moon
06	16:32	FIRST QUARTER MOON
08	06:00	Mercury at greatest elong: 18.2°E
09	00:27	Mars (mag. −0.1) 1.7°S of Moon
09	11:06	Pollux 2.0°N of Moon
12	06:07	Regulus 2.2°S of Moon
14	06:55	FULL MOON
14	06:59	Total Lunar Eclipse; mag. 1.178
14		γ-Normid Meteor Shower maximum
16	19:16	Spica 0.3°N of Moon
17	16:37	Moon at apogee: 405,754 km
20	09:02	Autumnal Equinox
20	15:58	Antares 0.5°N of Moon
22	11:29	LAST QUARTER MOON
29	10:47	Partial Solar Eclipse; mag.=0.938
29	10:58	NEW MOON
29	19:29	Mars (mag. 0.4) 3.9°S of Pollux
29	05:26	Moon at perigee: 358,127 km

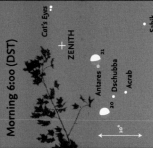

Evening 21:00 (DST)

9 March • In the northern sky, the Moon is in the constellation of Gemini, close to Mars (mag. −0.1), Pollux and Castor.

Morning 6:00 (DST)

20–21 March • The Moon is high in the north, when it passes Dschubba, (δ Sco) Acrab (β Sco) and Antares.

Evening 21:00 (DST)

5–7 March • In the northwest, the Moon passes the Pleiades, Aldebaran, Jupiter (now faded to mag. −2.3) and Enath (β Tau).

Morning 6:00 (DST)

17 March • High in the west, the waning gibbous Moon is close to Spica.

Evening 21:00 (DST)

12 March • The Moon is between Regulus and Algieba (γ Leo), in the constellation of Leo.

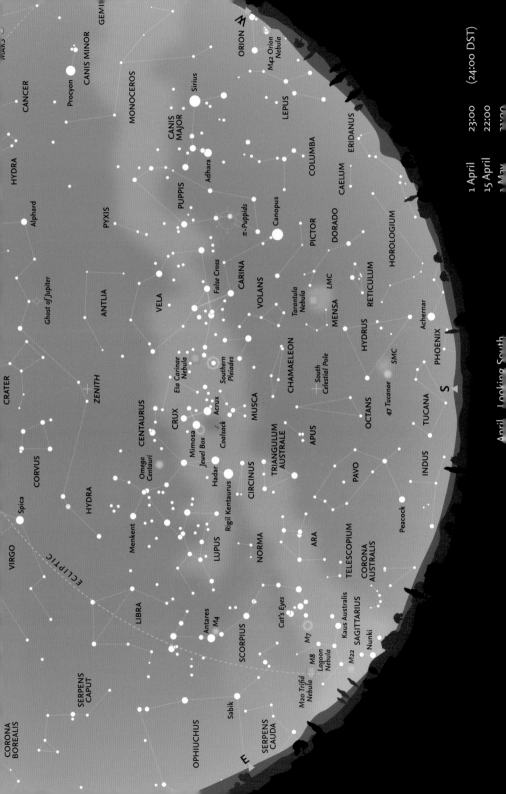

April, Looking South

1 April   23:00   (24:00 DST)
15 April   22:00
1 May   21:00

April – Looking South

Daylight Saving Time ends where used in Australia and in New Zealand on Sunday 6 April with the arrival of autumn. **Crux** is now high in the south, with the two brightest stars of **Centaurus** (α and β Centauri, **Rigil Kentaurus** and **Hadar**, respectively) to its east. The magnificent globular cluster, **Omega Centauri**, is readily visible high in the sky. West of Crux, both the **Southern Pleiades** and the **Eta Carinae Nebula** are clearly seen, two-thirds of the way towards the zenith. Farther west, the **False Cross** is beginning to decline towards the horizon. **Canopus** (α Carinae) is even lower, and **Orion** has now disappeared below the horizon. **Canis Major** and brilliant **Sirius** are also descending in the west. **Achernar** (α Eridani) is skimming the southern horizon, but **Peacock** (α Pavonis) is now slightly higher and more easily visible. Although the **LMC** is roughly as high as the South Celestial Pole, the **SMC** and **47 Tucanae** are rather low (but still visible) in the south. In the east, the whole of **Scorpius** is now well clear of the horizon with the inconspicuous constellation of **Libra** preceding it along the ecliptic. The dense regions of the Milky Way in **Sagittarius** become visible later in the night.

Meteors

Two meteor showers, in particular, occur in April. The **π-Puppid** shower begins on 15 April, and conditions are moderately favourable in 2025 with the shower maximum on 23–24 April, when the Moon is a waning crescent. The hourly rate is variable but the meteors tend to be faint. The parent body is the comet 26P/Grigg-Skjellerup. A second, more prolific, shower, the **η-Aquariids**, begins on 19 April, when the Moon is waning gibbous, and continues into May.

Omega Centauri (NGC 5139) is the largest and finest globular cluster in the sky (and in the Milky Way galaxy). It is believed to contain 10 million stars and differs in chemical composition and nature so greatly from other globulars that it may be the core of a disrupted dwarf galaxy, captured by the Milky Way.

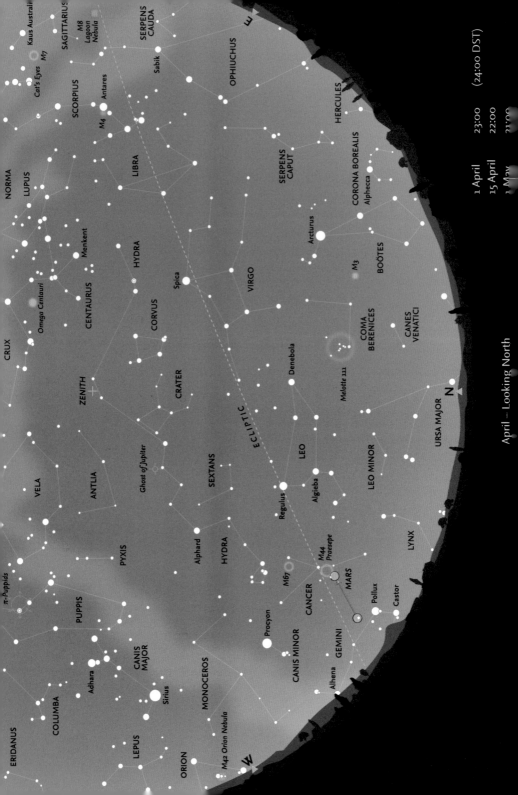

April – Looking North

1 April   23:00   (24:00 DST)
15 April   22:00
1 May   21:00

April – Looking North

Leo is the most prominent constellation in the northern sky in April. **Gemini**, with **Castor** and **Pollux**, is low on the horizon in the west, and **Cancer** lies between the two constellations. To the east of Leo, the whole of **Virgo**, with **Spica** (α Virginis) its brightest star, is clearly visible. The constellation of **Libra** is farther east along the ecliptic. Above Leo and Virgo, the complete length of **Hydra** is visible, **Alphard** (α Hydrae) forms a prominent triangle with **Regulus** and **Procyon** in **Canis Minor**. High in the sky, the small constellations of **Sextans** and **Crater**, together with the rather brighter **Corvus**, lie between Leo, Virgo and Hydra.

Boötes and **Arcturus** are prominent in the northeastern sky, together with the globular cluster M3 in **Canes Venatici**. The circlet of **Corona Borealis** is close to the horizon. Between Leo and Boötes lies the constellation of **Coma Berenices**, notable for being the location of the open cluster Melotte 111 (this is sometimes called the Coma Cluster and can be confused with the Coma Cluster of galaxies (Abell 1656) east of **Denebola** in **Leo**). There are about 1,000 galaxies in Abell 1656, shown on the chart on page 61, and which is located near the North Galactic Pole, where we are looking out of the plane of the Galaxy and are thus able to see deep into space. Only about ten of the brightest galaxies in the Coma Cluster are visible with the largest amateur telescopes.

A very large, and frequently ignored, open star cluster, Melotte 111, also known as the Coma Cluster, is readily visible in the northen sky during April and May (south is up).

The Moon's phases for April 2025

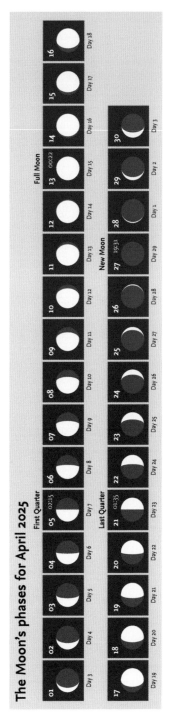

April – Moon and Planets

The Moon

On 3 April the waxing crescent Moon will be visible 5.5°N of **Jupiter** (mag. -2.1). On 5 April the First Quarter Moon will pass 2.2°N of **Mars** (mag. 0.5) and 2.0°S of **Pollux**. Three days later the bright waxing gibbous Moon will lie 2.0°N of **Regulus**. The Full Moon on 13 April will be visible less than half a degree from **Spica**. On 16 April the Moon and **Antares** are 0.4° apart. Last Quarter occurs on 21 April. On 25 April the waning crescent Moon passes 2.4°S of **Venus** (mag. -4.5) and 2.3°N of **Saturn** (mag. 1.0), all are visible very low in the sky just before sunrise. A day later the faint Moon will lie 4.4°N of **Mercury** (mag. 0.2). The following day it is a New Moon. On 30 April the thin crescent Moon will appear 5.4°N of **Jupiter** (mag. -2.0) in the early evening sky.

The planets

On 11 April **Mercury** reaches its highest point in the sky, at mag. 0.3 during the daytime. On 21 April it will have greatest western elongation. On 24 April **Venus** will reach greatest brightness (mag. -4.5), visible shortly before sunrise – it will rise earlier as the month progresses. **Mars** (mag. 0.4 to 0.9) is high in the sky several hours before midnight. **Jupiter** (mag. -2.1 to -2.0) in **Taurus** is also visible in the early evening. **Saturn** (mag. 1.2) will be close to **Venus** and **Neptune** on 28 April – they will all pass within 4° of each other in **Pisces** before sunset. **Uranus** (mag. 5.8) is in **Taurus** and **Neptune** (mag. 7.8) is in **Pisces.**

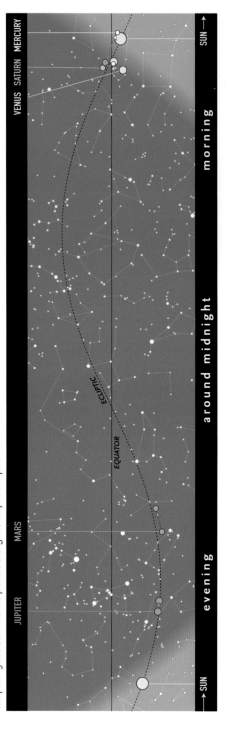

The path of the Sun and the planets along the ecliptic in April.

Calendar for April

01	20:28	Pleiades 0.6°S of Moon
03	00:23	Jupiter (mag. −2.1) 5.5°S of Moon
05	02:15	FIRST QUARTER MOON
05	16:46	Pollux 2.1°N of Moon
05	19:04	Mars (mag. 0.5) 2.0°S of Moon
08	11:51	Regulus 2.4°S of Moon
10	12:00	Mercury (mag. 1.0) 2.1°N of Saturn
13		FULL MOON
13	00:22	Spica 0.3°N of Moon
13	22:48	Moon at apogee: 406,295 km
16	22:00	Mars at aphelion: 1.66606 AU
16	22:19	Antares 0.4°N of Moon
21	01:35	LAST QUARTER MOON
21	19:00	Mercury at greatest elong: 27.4°W
23		π-Puppid Meteor Shower maximum
25	01:21	Venus (mag. −4.5) 2.4°N of Moon
25	04:15	Saturn (mag. 1.0) 2.3°S of Moon
26	01:05	Mercury (mag. 0.2) 4.4°S of Moon
27	16:15	Moon at perigee: 357,119 km
27	19:31	NEW MOON
28	19:00	Venus (mag. −4.8) 3.7°N of Saturn (mag. 1.2)
29	06:35	Pleiades 0.5°S of Moon
30	17:33	Jupiter (mag. −2.0) 5.4°S of Moon

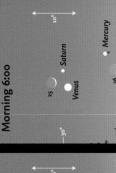

Evening 19:30 (DST)

Pleiades · Aldebaran · Jupiter · Elnath · Capella · Menkalinan

1–3 April • The crescent Moon passes the Pleiades, Aldebaran, Jupiter (mag. −2.1) and Elnath (β Tau). Capella (α Aur) and Menkalinan (β Aur) are close to the horizon.

Evening 23:00 (DST)

Procyon · Mars · Pollux · Castor · Moon · Alhena

5 April • The Moon passes Mars (mag. 0.5). Castor and Pollux. Alhena (γ Gem) and Procyon are close by.

Evening 19:00

Spica

12–13 April • The Moon is nearly Full when it passes close to Spica, the brightest star in Virgo.

Evening 21:00

Dschubba · Acrab · Antares

16–17 April • The Moon is in the constellation of Scorpius and passes Dschubba, Acrab and Antares.

Morning 6:00

Saturn · Venus · Mercury

25–26 April • The crescent Moon moves between Venus and Saturn. On 26 April it is close to Mercury.

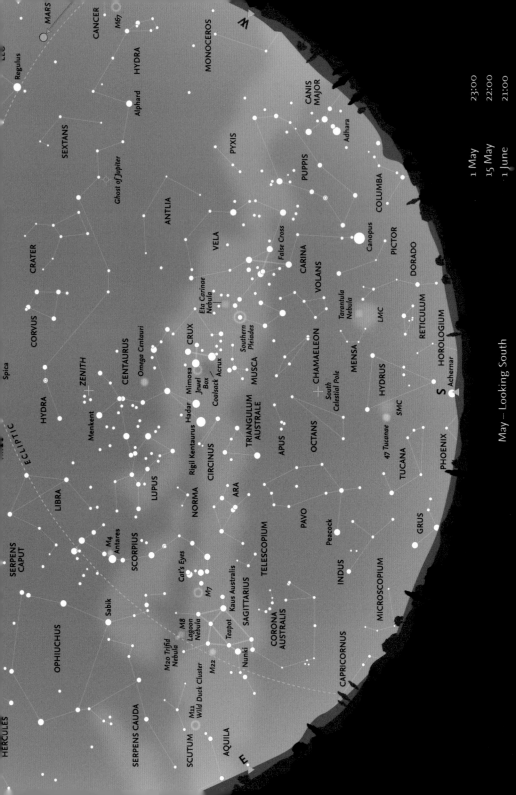

May — Looking South

1 May   23:00
15 May   22:00
1 June   21:00

May – Looking South

In the west, **Canis Major** has set below the horizon, and **Canopus** (α Carinae) and **Puppis** are getting rather low. The whole of **Sagittarius** is now clearly seen in the east, with **Corona Australis** beneath it and **Scorpius** higher above. **Centaurus** is fully seen high in the south, with **Crux**, the **Eta Carinae Nebula**, the **Southern Pleiades** and the **False Cross** with **Vela** along the Milky Way to the west. In the south, **Achernar** (α Eridani) is skimming the southern horizon and the **Small Magellanic Cloud** (SMC) is beginning to rise higher in the sky, unlike the **Large Magellanic Cloud** (LMC) which is now lower. **Pavo** and **Peacock** (α Pavonis) are now much higher and even the faint constellations of **Tucana** and **Indus** are visible between Pavo and the southwestern horizon.

Meteors

The **η-Aquariids** are one of the two meteor showers associated with Comet 1P/Halley (the other being the **Orionids**, in October). The radiant is near the celestial equator, close to the 'Water Jar' in **Aquarius**, well below the horizon. However, meteors may still be seen in the eastern sky even when the radiant is below the horizon. There is a radiant map for the η-Aquariids on page 30. Their maximum in 2025, on 5 May, occurs a day after the First Quarter Moon. Conditions will be more favourable shortly after midnight once the Moon has set. Maximum hourly rate is about 40 per hour and a large proportion (about 25 per cent) of the meteors leave persistent trains.

Some faint constellations around the South Celestial Pole (SCP). Below the Pole, a part of the constellation of Octans is visible. Then, a little lower, Apus and, farther down, Triangulum Australe. To the left is Musca and just above the centre of the image is Chamaeleon. The bright star near the top edge of the image is Miaplacidus, in the constellation of Carina.

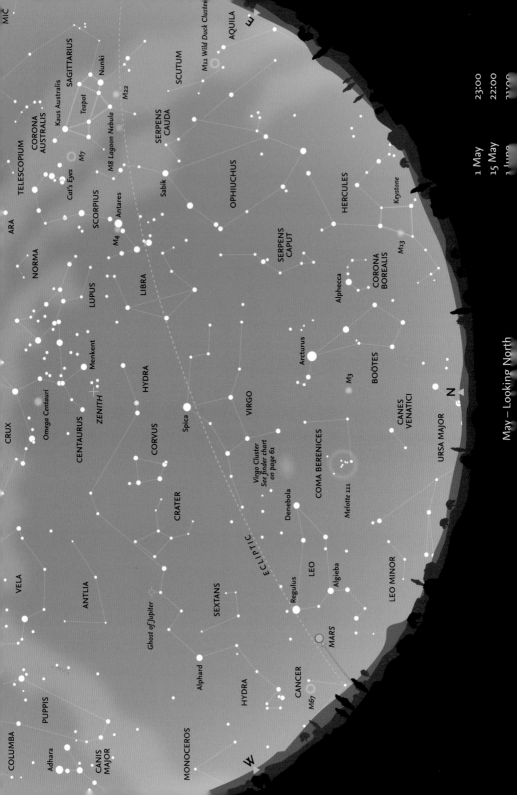

May – Looking North

May – Looking North

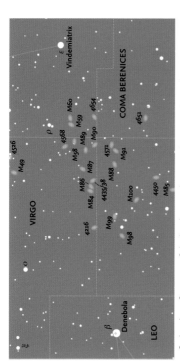

Boötes is almost due north, with brilliant, slightly orange-coloured *Arcturus* extemely prominent. The distinctive circlet of *Corona Borealis* is clearly visible to its east. The brightest star (α Coronae Borealis) is known as *Alphecca*. *Hercules* is rising in the east and becomes clearly visible later in the night. To the west of Boötes is the inconspicuous constellation of *Coma Berenices*, with the cluster *Melote 111* (see page 55) and, above it, the Virgo Cluster of galaxies (see here).

The large constellation of *Ophiuchus* (which actually crosses the ecliptic, and is thus the 'thirteenth' zodiacal constellation) is climbing into the eastern sky. Before the constellation boundaries were formally adopted by the International Astronomical Union in 1930, the southern region of Ophiuchus was regarded as forming part of the constellation of *Scorpius*, which had been part of the zodiac since antiquity.

Early in the night, the constellation of *Virgo*, with *Spica* (α Virginis), lies due north, with the rather faint constellation of *Libra* to its east. Farther along the ecliptic are Scorpius and brilliant, reddish *Antares* (α Scorpii). In the west, the constellation of *Leo* is readily visible now,

A finder chart for some of the brightest galaxies in the Virgo Cluster (see page 60). All stars brighter than magnitude 8.5 are shown (south is up).

and both *Regulus* and *Denebola* (α and β Leonis, respectively) are prominent.

Virgo contains the nearest large cluster of galaxies, which is the centre of the Local Supercluster, of which the Milky Way forms part. The Virgo Cluster contains some 2,000 galaxies, the brightest of which are visible in amateur telescopes.

The Moon's phases for May 2025

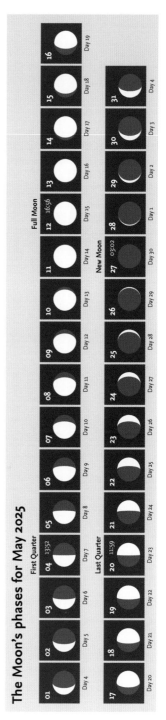

May – Moon and Planets

The Moon

On 3 May the Moon will be 2.1°S of *Pollux* around midnight, and the following night the Moon and Mars (mag. 1.0) will be 2.1° apart. On 5 May the Moon and *Regulus* are 2.0° apart. On 10 May *Spica* is 0.4°N of the waxing gibbous Moon. The Full Moon is on 12 May. On 14 May the Moon will be 0.3°S of *Antares*. The Last Quarter Moon is on 20 May. On 22 May the waning crescent Moon will be almost 3°N of *Saturn* (mag. 0.9) during the daytime. On 24 May the Moon and *Venus* (mag. -4.4) will be 4° apart. The New Moon will be visible on 27 May. On 30 May the thin crescent Moon will lie 2.3° from *Pollux*.

The planets

On 2 May minor planet (4) *Vesta* will be at opposition in *Libra*, reaching its highest point at midnight. *Mercury* (mag. 0.1 to -2.3) climbs above the horizon shortly before sunrise. *Venus* is also visible at dawn, on 31 May it will reach greatest western elongation at mag. -4.4. *Mars* (mag. 0.9 to 1.2) continues to move eastward. *Jupiter* (mag. -2.0 to -1.9) in *Taurus* sets earlier in the evening as the month progresses. *Saturn* (mag. 1.2 to 1.1) in *Pisces* is above the horizon at dawn. *Uranus* (mag. 5.8) follows in close proximity to the Sun in the sky. *Neptune* (mag. 7.8) is in *Pisces*.

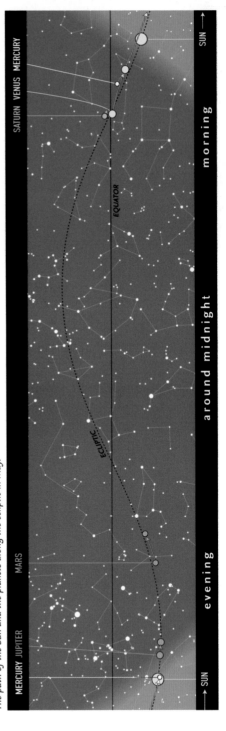

The path of the Sun and the planets along the ecliptic in May.

Calendar for May

02	06:28	Minor planet (4) Vesta at opposition (mag. 5.7)
03	00:02	Pollux 2.2°N of Moon
03	23:12	Mars (mag. 1.0) 2.1°S of Moon
04	13:52	FIRST QUARTER MOON
05		η-Aquariid Meteor Shower maximum
05	17:58	Regulus 2.2°S of Moon
10	07:43	Spica 0.4°N of Moon
11	00:49	Moon at apogee: 406,245 km
12	16:56	FULL MOON
14	04:10	Antares 0.3°N of Moon
20	11:59	LAST QUARTER MOON
22	17:51	Saturn (mag. 0.9) 2.8°S of Moon
23	23:52	Venus (mag. -4.4) 4.0°S of Moon
26	01:37	Moon at perigee: 359,023 km
27	03:02	NEW MOON
28	13:12	Jupiter (mag. -1.9) 5.2°S of Moon
30	09:13	Pollux 2.4°N of Moon
31	13:00	Mercury at perihelion

Evening 19:00

2–4 May • The Moon passes Castor, Pollux and Mars (mag. 0.9). After First Quarter (on 4 May) it moves towards Regulus and Algieba, in the constellation of Leo, the Lion.

Early morning 3:00

10–11 May • Due west, the Moon passes close to Spica in the constellation of Virgo.

Early morning 3:00

14–15 May • The Moon passes Dschubba, Acrab and Antares. Sabik and the 'Cat's Eyes' are nearby.

Early morning 4:00

23–24 May • In the early morning, the Moon passes Saturn (mag. 1.1) and brilliant Venus (mag. -4.5).

Evening 18:00

30 May • Shortly after dawn, the waxing crescent Moon is close to Pollux and Castor.

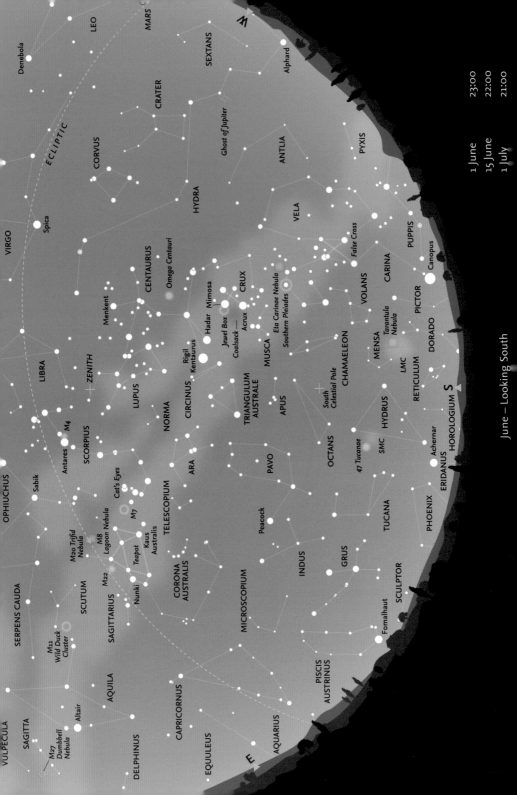

June – Looking South

1 June   23:00
15 June   22:00
1 July   21:00

June – Looking South

Both **Canopus** (α Carinae) and **Achernar** (α Eridani) are skimming the southern horizon. Although the whole of **Carina** is visible, all of **Eridanus** (except **Achernar**) is hidden below the horizon. The **Large Magellanic Cloud** (LMC) is low, although the **Small Magellanic Cloud** (SMC), the globular cluster, **47 Tucanae** and **Hydrus** are now rather higher. **Alphard** (α Hydrae) is on the horizon, and the constellation of **Sextans** is becoming low. However, the remainder of the long constellation of Hydra is clearly seen as are the two constellations of **Crater** and **Corvus** to its north. The **False Cross** between Carina and **Vela** is beginning to descend in the west, but Vela itself is clearly visible. The whole of both **Crux** and **Centaurus** are clearly seen, as are the magnificent globular cluster, **Omega Centauri**, and the constellation of **Lupus**, closer to the zenith. The constellation of **Triangulum Australe** is on the meridian, roughly halfway between the South Celestial Pole and the zenith. Both **Scorpius**, with brilliant, red **Antares** (α Scorpii) and **Sagittarius** are high overhead, with the faint constellation of **Corona Australis** visible below them. The whole of **Capricornus** is visible and the constellation of **Grus** has now risen above the horizon with the faint constellation of **Indus** between it and **Pavo**. To the east of Grus is **Piscis Austrinus**, although brilliant **Fomalhaut** (α Piscis Austrini) is only just clear of the horizon, and becomes clearly visible only later in the night and later in the month.

The distinctive shape of the zodiacal constellation of Scorpius. The red supergiant star, Antares is close to the zenith in June. The bright spot near the left side of the image is the open cluster M7, also known as Ptolomy's Cluster (north is up).

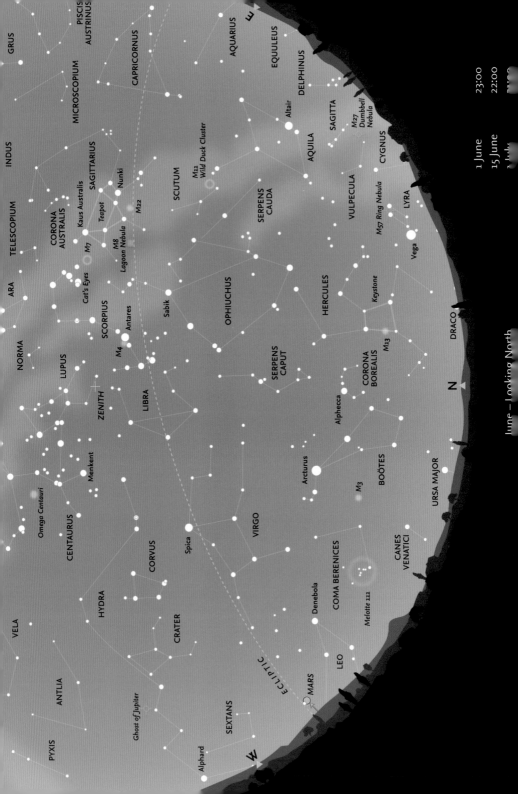

June – Looking North

1 June   23:00
15 June   22:00

June – Looking North

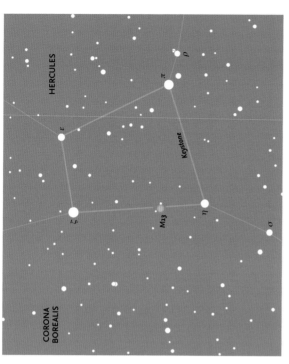

Vega in *Lyra* is just above the northeastern horizon. Closer to the meridian, the whole of *Hercules* is visible including the **'Keystone'** and **M13**, widely regarded as the finest globular cluster in the northern hemisphere. The small constellation of *Corona Borealis* is almost due north. To the west of the meridian is *Boötes* and *Arcturus* (α Boötis). Much of *Leo* is now below the western horizon, but *Denebola* (β Leonis) is still visible. Higher in the sky, the whole of *Virgo* is clearly visible and, still higher, not far from the zenith is the constellation of *Libra*. To its east is *Scorpius* and reddish *Antares*. Farther along the ecliptic, both the constellations of *Sagittarius* and *Capricornus* are completely visible. Between Hercules and Sagittarius is the large constellation of *Ophiuchus* and, to its east, *Aquila* and *Altair* (α Aquilae), one of the stars of the (northern) Summer Triangle. Another of the three stars, *Vega* in *Lyra*, is skimming the northern horizon.

Finder chart for M13, the finest globular cluster in the northern sky. All stars down to magnitude 7.5 are shown (south is up).

The Moon's phases for June 2025

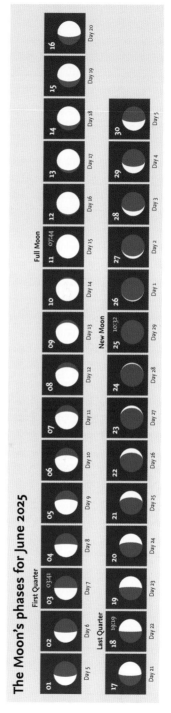

June – Moon and Planets

The Moon

On 1 June, the waxing crescent Moon will be 1.4°N of *Mars* (mag. 1.2). The following day the Moon will be less than 2° from *Regulus*. On 3 June the Moon is First Quarter. Three days later the waxing gibbous Moon is 0.5°S of *Spica*. On 10 June the Moon passes 0.3°S of *Antares*, a day later it is a Full Moon. On 18 June the Moon is at Last Quarter. On 19 June the waning Moon, *Saturn* (mag. 1.0) and *Neptune* (mag. 7.8) are all within 3–4° of each other, above the horizon in the dawn sky. On 25 June there will be a New Moon. On 27 June the thin crescent Moon will pass less than 3°N of *Mercury* (mag. 0.1), visible shortly after sunset. Two days later the Moon and *Regulus* are 1.5° apart. Shortly after midnight on 30 June the Moon and *Mars* (mag. 1.4) will be 0.2°

apart. The lunar occultation of Mars will be visible from some parts of Peru, Equador and Colombia.

The planets

Mercury (mag. -2 to 0.3) sets with the Sun in the early part of the month, increasing its separation towards the end of the month as it approaches eastern elongation. *Venus* (mag. -4.4 to -4.2) climbs above the eastern horizon in *Aries* before sunrise. *Mars* (mag. 1.2 to 1.4) in *Leo* is visible in the evening sky, setting earlier towards the end of the month. *Jupiter* (mag. -1.9) moves from *Taurus* to *Gemini* on 12 June and sets shortly after the Sun. *Saturn* (mag. 1.0) and *Neptune* (mag. 7.8) will pass within a degree of each other on 29 June in *Pisces*. *Uranus* (mag. 5.8) is in *Taurus* in the dawn sky.

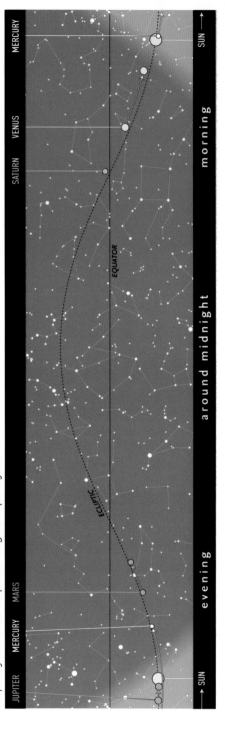

The path of the Sun and the planets along the ecliptic in June.

Calendar for June

01	02:00	Venus at greatest elong: 45.9°W
01	09:49	Mars (mag. 1.2) 1.4°S of Moon
02	01:30	Regulus 1.8°S of Moon
03	03:41	FIRST QUARTER MOON
06	14:15	Spica 0.5°N of Moon
07	10:42	Moon at apogee: 405,553 km
10	10:25	Antares 0.3°N of Moon
11	07:44	FULL MOON
12	03:00	Venus at aphelion
17	02:05	Mars (mag. 1.4) 0.7°N of Regulus
18	19:19	LAST QUARTER MOON
19	03:47	Saturn (mag. 1.0) 3.4°S of Moon
21	02:42	Summer Solstice
21	19:51	Mercury (mag. -0.2) 4.8°S of Pollux
23	02:59	Pleiades 0.6°S of Moon
23	04:43	Moon at perigee: 363,178 km
25	19:19	Jupiter (mag. -1.9) 5.1°S of Moon
25	10:32	NEW MOON
26	19:14	Pollux 2.5°N of Moon
27	06:02	Mercury (mag. 0.1) 2.9°S of Moon
29	10:26	Regulus 1.5°S of Moon

Evening 21:00

1–2 June • In the northwest, the Moon passes Mars (mag. 1.2), Regulus and Algieba.

Evening 19:30

10 June • The Moon is very close to Antares, in the constellation of Scorpius.

Evening 21:00

16 June • Mars is very close to Regulus, and almost equally as bright.

After midnight 2:00

19–20 June • The Moon passes Saturn (mag. 1.0) in the east-northeastern sky, halfway between Diphda (β Cet) and Markab (α Peg)

Evening 19:00

29–30 June • The Moon is in the constellation of Leo, passing Regulus and Mars, Algieba, Denebola (β Leo) and Alphard (α Hya) are all nearby.

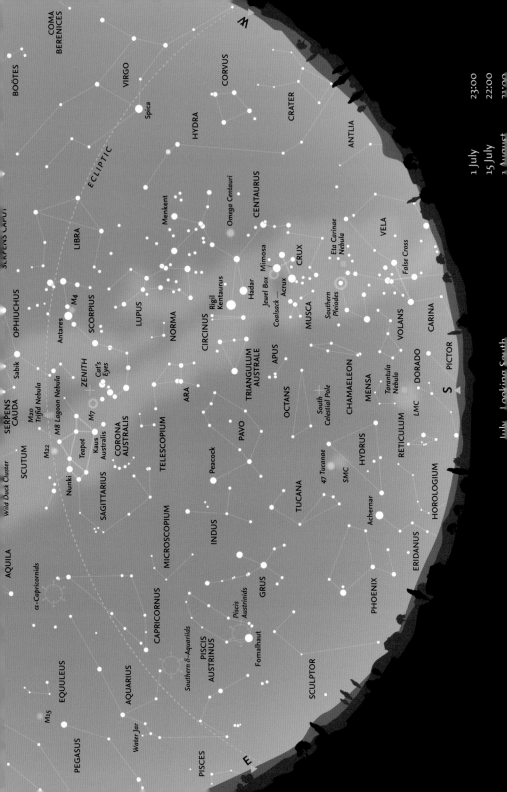

July – Looking South

1 July   23:00
15 July   22:00
1 August   21:00

July – Looking South

Part of the central Milky Way in the constellations of Norma and Scorpius, showing the dark clouds of obscuring dust (north is up).

Although the **Large Magellanic Cloud** (LMC) is almost due south, it is very low. Slightly farther west, the **False Cross** is nearing the horizon. Some of **Hydra** remains visible, but the constellation of **Crater** is becoming very low. **Corvus** is still easy to see. The zodiacal constellation of **Virgo** will soon be disappearing in the west. **Crux**, **Centaurus** and **Lupus** are still clearly seen, high in the sky, as is **Scorpius**, the tail of which is near the zenith. **Achernar** (α Eridani), **Hydrus** and the **Small Magellanic Cloud** (SMC) are now higher above the horizon and easier to observe. The whole of the constellation of **Phoenix** is also now clear of the horizon. Above it are **Tucana** and, halfway to the zenith, the constellation of **Pavo**. **Grus** and **Pisces Austrinus** (with brilliant **Fomalhaut**) are now fully visible. Higher in the sky are the zodiacal constellations of **Capricornus** and **Aquarius**, and even the westernmost portion of **Pisces** is rising above the horizon.

Meteors

July brings increasing meteor activity, mainly because there are several minor radiants active in the constellations of **Capricornus** and **Aquarius**. The first shower, the **α-Capricornids**, active from 3 July to 15 August (peaking 30 July), does often produce very bright fireballs. The maximum rate, however, is only about 5 per hour. The parent body is Comet 169/NEAT. The most prominent shower is probably that of the **Southern δ-Aquariids**, which are active from around 12 July to 23 August, also with a peak on 30 July, although even then the hourly rate is unlikely to reach 25 meteors per hour. In this case, the parent body is possibly Comet 96P/Machholz. This year, both showers occur when the Moon is a waxing crescent a few days before First Quarter, so conditions are most favourable after midnight once the Moon has set. The **Piscis Austrinids** begin on 15 July and continue until 10 August. Maximum is on 28 July, but the rate is only about 5 per hour. The **Perseids** begin on 17 July and peak on 12 August.

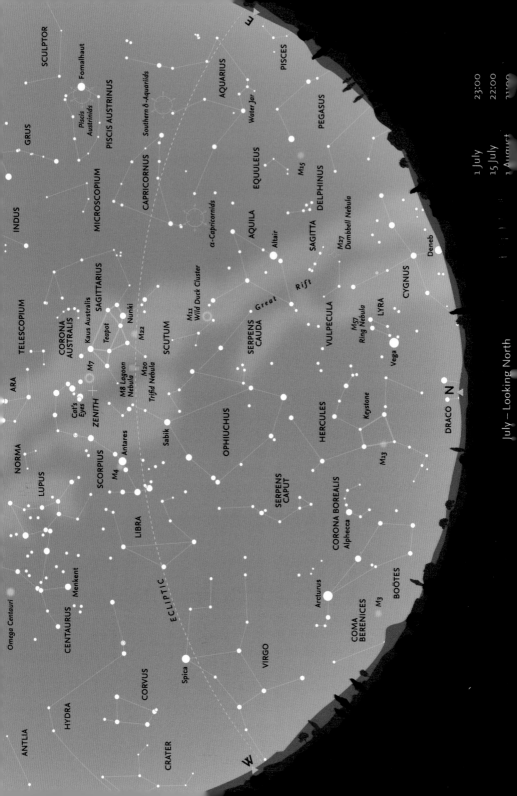

July – Looking North

CRATER
ANTLIA
HYDRA
CORVUS
Spica
VIRGO
ECLIPTIC
Menkent
CENTAURUS
Omega Centauri
LUPUS
NORMA
LIBRA
M4
SCORPIUS
Antares
Sabik
ARA
ZENITH
Cat's
Eyes
M7
CORONA
AUSTRALIS
TELESCOPIUM
SAGITTARIUS
Kaus Australis
Teapot
Nunki
M22
M8 Lagoon
Nebula
M20
Trifid Nebula
OPHIUCHUS
SCUTUM
M11
Wild Duck Cluster
SERPENS
CAUDA
Great Rift
MICROSCOPIUM
CAPRICORNUS
α-Capricornids
INDUS
GRUS
PISCIS AUSTRINUS
Piscis
Austrinids
Fomalhaut
SCULPTOR
Southern δ-Aquariids
AQUARIUS
Water Jar
PISCES
PEGASUS
EQUULEUS
M15
SAGITTA
DELPHINUS
AQUILA
Altair
M27
Dumbbell Nebula
Deneb
CYGNUS
VULPECULA
M57
Ring Nebula
LYRA
Vega
DRACO N
HERCULES
Keystone
M13
SERPENS
CAPUT
CORONA BOREALIS
Alphecca
BOÖTES
Arcturus
M3
COMA
BERENICES

E

W

1 July   23:00
15 July   22:00
1 August

July – Looking North

The constellations of **Hercules** and **Lyra** are on either side of the meridian and both are clearly visible well above the horizon. To the east, **Deneb** (α Cygni) is skimming the horizon, but nearly all of the rest of the constellation is easily seen, as is **Aquila** with **Altair** (α Aquilae). Between **Cygnus** and Aquila lies the small constellation of **Sagitta**, with **M27** (the Dumbbell Nebula, a planetary nebula). Even farther east, the whole extent of both the zodiacal constellations of **Aquarius** and **Capricornus** is visible. Above Aquila, farther along the **Great Rift** is the small constellation of **Scutum**, with **M11**, the Wild Duck Cluster. To its east is the tiny, but distinctive constellation of **Delphinus**. Still farther along the Great Rift lies the centre of the Milky Way Galaxy (in **Sagittarius**) and, near it, two emission nebulae: **M8** (the Lagoon Nebula) and **M20** (the Trifid Nebula). In the western sky, the constellation of **Boötes** and **Arcturus** (α Boötis) are beginning to approach the horizon, but **Corona Borealis** is still clearly seen. Even farther west, the whole of **Virgo** is visible, with **Libra** above it. The large constellation of **Ophiuchus** and the two halves of **Serpens** lie between Hercules and the zenith. **Scorpius** is draped around the actual zenith with Sagittarius to its east.

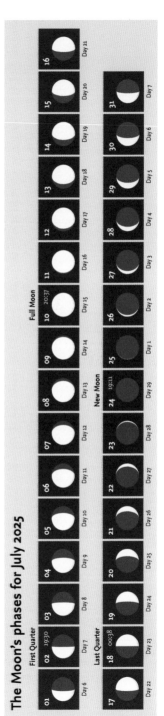

The constellation of Virgo. Spica (α Virginis) is the bright blue star near the top of the image, with Mars to its northeast (south is up).

The Moon's phases for July 2025

First Quarter

01	02 19:30	03	04	05	06	07	08
Day 6	Day 7	Day 8	Day 9	Day 10	Day 11	Day 12	Day 13

Full Moon

09	10 20:37	11	12	13	14	15	16
Day 14	Day 15	Day 16	Day 17	Day 18	Day 19	Day 20	Day 21

Last Quarter

17	18 00:38	19	20	21
Day 22	Day 23	Day 24	Day 25	Day 7

New Moon

23	24 19:11	25	26	27	28	29	30	31
Day 27	Day 28	Day 1	Day 2	Day 3	Day 4	Day 5	Day 6	Day 7

22
Day 26

July – Moon and Planets

The Moon

A day after the First Quarter Moon *Spica* is 0.8°N of the Moon. On 7 July the Moon is close to *Antares* at only 0.4° separation. Full Moon is on 10 July. On 16 July the Moon will pass less than 4°N of *Saturn* (mag. 0.9), *Neptune* will also be close by. The Last Quarter Moon is on 18 July. On 23 July the waning Moon will lie within 5° of *Jupiter* (mag. -1.9), visible at dawn. Two days after New Moon on 26 July the crescent Moon will be 1.4°N of *Regulus*. On 28 July the waxing crescent Moon will be just over 1°S of *Mars* (mag. 1.6). On 31 July *Spica* and the Moon are close together again in the sky, just over a degree apart.

The planets

Mercury will reach greatest western elongation on 4 July, bright at magnitude 0.5 in the early evening sky, dimming to mag. 5.0 in late summer. *Venus* (mag. -4.1 to -4.0) continues to be the morning star placed in *Taurus*, it moves eastwards towards *Orion* at the end of the month. *Mars* is visible in the evening sky in *Leo*, moving into *Virgo* on 28 July. *Jupiter* (mag. -1.9) rises with the Sun in *Gemini*. *Saturn* (mag. 0.9) enters retrograde motion in *Pisces* on 13 July. *Uranus* (mag. 5.8) is in *Taurus*. *Neptune* (mag. 7.8) will enter retrograde motion on 4 July in *Pisces*.

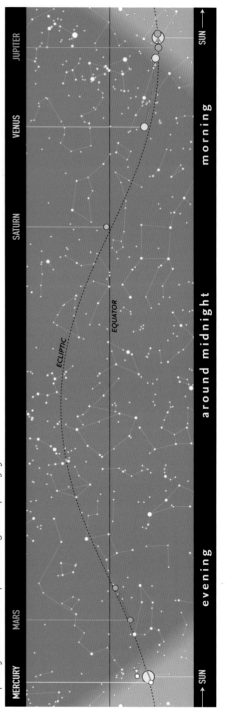

The path of the Sun and the planets along the ecliptic in July.

Calendar for July

02	19:30	FIRST QUARTER MOON
03	21:00	Earth at aphelion: 1.01664 AU
03	21:39	Spica 0.8°N of Moon
04	04:00	Mercury at greatest elong: 25.9°E
05	02:29	Moon at apogee: 404,627 km
07	17:37	Antares 0.4°N of Moon
10	20:37	FULL MOON
13	08:32	Venus (mag. -4.1) 3.1°N of Aldebaran
14	13:00	Mercury at aphelion
16	10:19	Saturn (mag. 0.9) 3.8°S of Moon
18	00:38	LAST QUARTER MOON
20	10:27	Pleiades 0.7°S of Moon
20	13:52	Moon at perigee: 368,047 km
23	04:20	Jupiter (mag. -1.9) 4.9°S of Moon
24	19:11	NEW MOON
26	19:44	Regulus 1.4°S of Moon
28	19:45	Mars (mag. 1.6) 1.3°N of Moon
30		α-Capricornid Meteor Shower maximum
30		Southern δ-Aquariid Meteor Shower maximum
31	05:45	Spica 1.0°N of Moon

Evening 23:00

3-4 July • High in the west, the waxing gibbous Moon passes Spica, in the constellation of Virgo.

Early morning 3:00

8 July • The Moon is close to Antares. Dschubba and Acrab are closer to the horizon.

Morning 5:30

13 July • Venus (mag. -4.1) is close to Aldebaran, with the Pleiades farther north.

Morning 6:30

23 July • The thin faint crescent Moon is low in the northeast, in the company of Jupiter (mag. -1.9). Venus (mag. -4.0) is higher in the northeast.

Evening 21:00

31 July • The waxing crescent Moon is near Spica again. Arcturus, the brightest star in the northern hemisphere, is farther northwest,

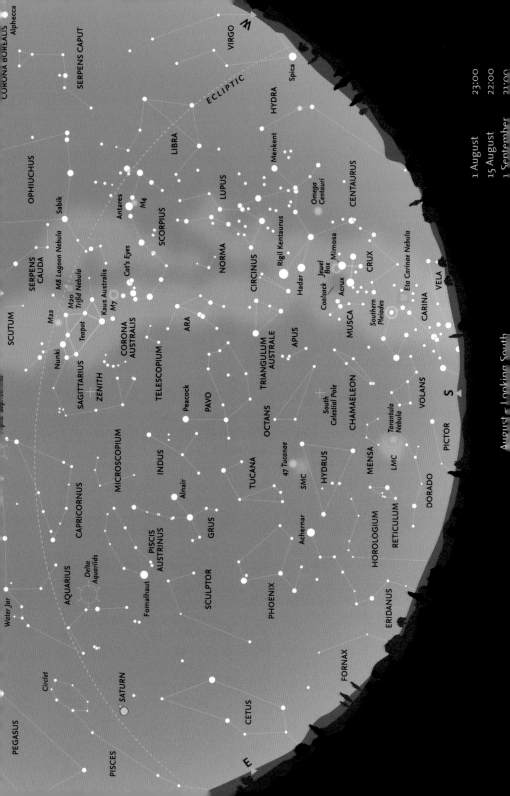

August - Looking South

1 August   23:00
15 August   22:00
1 September 21:00

August – Looking South

Three inconspicuous constellations: *Volans*, *Chamaeleon* and *Octans* are on the meridian, with *Pavo* higher towards the zenith. Much of *Carina* is below the horizon, although the *Eta Carinae Nebula* and the *Southern Pleiades* are still visible. The *Large Magellanic Cloud* (LMC) and the faint constellation of *Mensa* are slightly higher in the sky. *Crux* is now lower, but the whole of *Centaurus* and *Lupus* remains visible. Much of *Virgo* has set in the west and *Libra* is following it down towards the horizon. *Scorpius* and *Sagittarius* are still visible high in the sky. *Achernar*, the *Small Magellanic Cloud* (SMC) and *Hydrus* are now well clear of the horizon, but below them are more small, inconspicuous constellations: *Dorado*, *Reticulum* and *Horologium*. More of *Eridanus* is visible, together with parts of *Pictor* and *Fornax*. Much of *Cetus* has risen and the whole of the western arm of *Pisces* is now clearly seen. *Sculptor*, *Grus* and *Piscis Austrinus* lie halfway between the eastern horizon and the zenith, with faint *Microscopium* closer to the actual zenith.

Meteors

The *Piscis Austrinid* shower continues until about 10 August, after maximum on 28 July. The *Southern δ-Aquariids* reach maximum on 30 July, but are unlikely to exhibit more than about 25 meteors per hour. The *Perseids* peak on 12 August, when the rate may reach as high as 100 meteors per hour. These are best seen shortly before sunrise very close to the horizon. In 2025, maximum is three days after Full Moon, so conditions are not favourable, particularly at dawn. The Perseids are debris from Comet 109P/Swift-Tuttle (the Great Comet of 1862). Perseid meteors are fast and many of the brighter ones leave persistent trains. Some bright fireballs also occur during the shower. The *α-Aurigid* shower begins on 28 August and is a short shower, lasting until 5 September and reaching maximum in 2025 on 31 August.

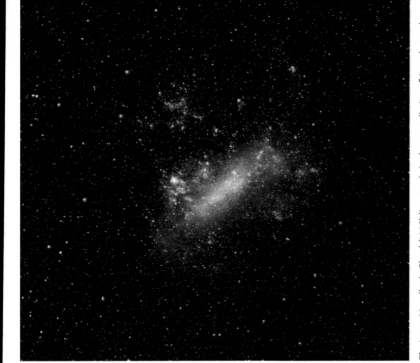

The *Large Magellanic Cloud* (LMC) lies mostly in the constellation of Dorado, partly extending into neighbouring Mensa. It contains many globular and open clusters as well as gaseous nebulae, such as the giant Tarantula Nebula.

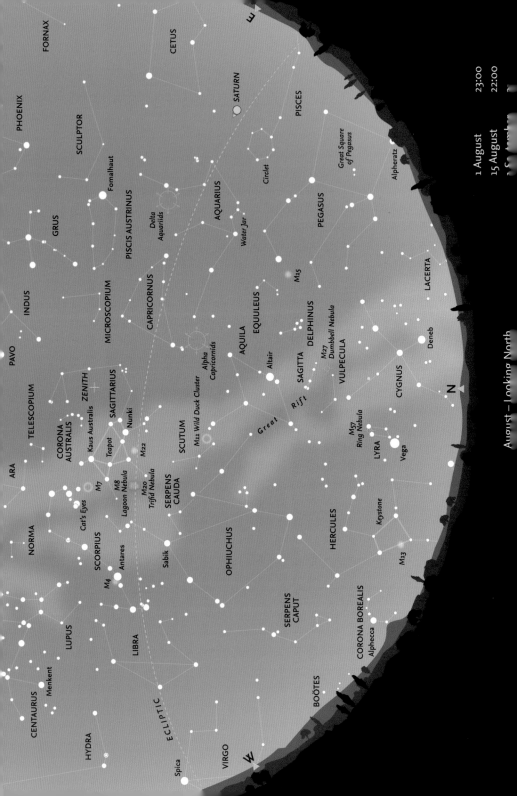

August – Looking North

August – Looking North

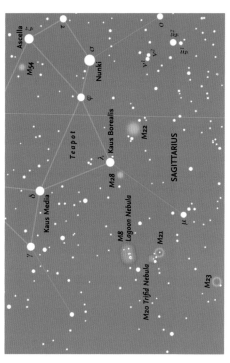

Cygnus, with brilliant **Deneb** (α Cygni), is now prominent just to the east of the meridian. The other two stars of the (northern) Summer Triangle, **Vega** in **Lyra** and **Altair** in **Aquila** are also unmistakeable in the sky. **Hercules,** however, is beginning to descend towards the northwestern horizon. Above it, the sprawling constellation of **Ophiuchus** is readily seen. The Great Square of **Pegasus** is now visible in the east, although **Alpheratz** (α Andromedae) is low on the horizon. The western side of **Pisces,** as well as **Aquarius** and **Capricornus,** is fully visible along the ecliptic. **Sagittarius** contains many gaseous nebulae, like **M8** (the Lagoon Nebula) and **M20** (the Trifid Nebula) and several globular clusters such as **M22.** The constellation is at the zenith, with Scorpius to the west.

A finder chart for the gaseous nebulae M8 (the Lagoon Nebula), M20 (the Trifid Nebula) and the globular cluster M22, all in Sagittarius. Clusters M21, M23 (open) and M28 (globular) are faint. The chart shows all stars brighter than magnitude 7.5 (south is up).

The Moon's phases for August 2025

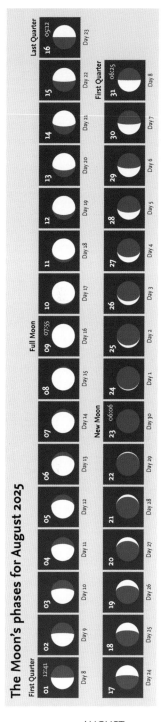

August – Moon and Planets

The Moon

On 1 August the Moon is First Quarter. The Moon passes south of *Antares* by 0.6° on 4 August, five days later it is full. On 12 August the Moon will be 4.0°N of *Saturn* (mag. 0.7). On 16 August the Moon is Last Quarter. Three days later the waning gibbous Moon will appear 4.8°N of *Jupiter* (mag. -2.0). The following day the Moon passes less than 5° from *Venus* (mag. -4.0) and 2.4°S of *Pollux*. On 21 August *Mercury* (mag. -0.3) will be high in the morning sky, 3.7°S of the thin crescent Moon in the dawn sky. Three days after the New Moon *Mars* (mag. 1.6) will appear 2.8°N of the waxing crescent Moon, setting shortly after the sunset. On 27 August the Moon and *Spica* are 1.1° apart and on 31 August the First Quarter Moon will be 0.7°S of *Antares*.

The Planets

Mercury (mag. -0.2) will be at greatest western elongation on 19 August. *Venus* (mag. -4.0) will be at its highest altitude in the morning sky. *Mars* (mag. 1.6) sets earlier in the evening as the month progresses. *Jupiter* (mag. -1.9 to -2.0) will be less than a degree from *Venus*, both in *Gemini*, on the morning of 12 August. *Saturn* (mag. 0.8 to 0.6) will be near *Neptune* (mag. 7.7) at dawn on 6 August in Pisces. *Uranus* (mag. 5.8 to 5.7) is in *Taurus*. Minor planet *(89) Julia* (mag. 8.5) is at opposition close to midnight on 10 August, and on the 26 August *(6) Hebe* (mag. 7.5) will be at opposition.

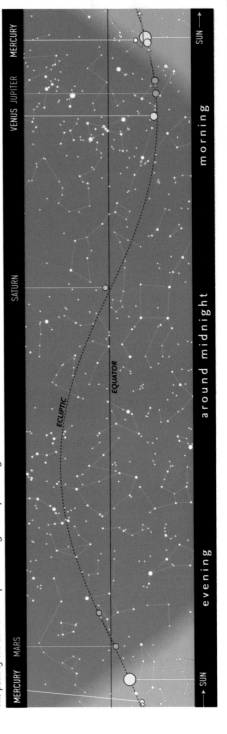

The path of the Sun and the planets along the ecliptic in August.

Calendar for August

01	12:41	FIRST QUARTER MOON
01	20:37	Moon at apogee: 404,164 km
04	01:40	Antares 0.6°N of Moon
09	07:55	FULL MOON
10	23:58	Minor planet (89) Julia at opposition (mag. 8.5)
12	07:00	Venus (mag. -4.0) 0.9°S of Jupiter
12	15:05	Saturn (mag. 0.7) 4.0°S of Moon
12		Perseid Meteor Shower maximum
14	18:01	Moon at perigee: 369,287 km
16	05:12	LAST QUARTER MOON
16	16:09	Pleiades 0.9°S of Moon
19	10:00	Mercury at greatest elong: 18.6°W
19	21:05	Jupiter (mag. -2.0) 4.8°S of Moon
20	12:07	Pollux 2.4°N of Moon
21	16:14	Mercury (mag. -0.3) 3.7°S of Moon
23	06:06	NEW MOON
26	13:53	Minor planet (6) Hebe at opposition (mag. 7.5)
26	16:41	Mars (mag. 1.6) 2.8°N of Moon
27	12:00	Mercury at perihelion
27	13:57	Spica 1.1°N of Moon
29	15:34	Moon at apogee: 404,552 km
31	06:25	FIRST QUARTER MOON
31	09:55	Antares 0.7°N of Moon
31		α-Aurigid Meteor Shower maximum

After midnight 1:00

4–5 August • The Moon passes Acrab (β Sco), Dschubba (δ Sco) and Antares. Sabik (η Oph) is farther west.

Morning 6:15

12 August • Venus (mag. -4.0) and Jupiter (mag. -1.9) are close together in the constellation of Gemini. Castor and Pollux are very low above the horizon.

Early morning 4:00

17–19 August • Shortly after Last Quarter the Moon passes the Pleiades and moves towards Elnath. Aldebaran is halfway between the Pleiades and Bellatrix (γ Ori).

Morning 6:00

20 August • The crescent Moon passes Jupiter and Venus. Castor, Pollux and also Mercury (mag. -0.1) are probably too low to detect.

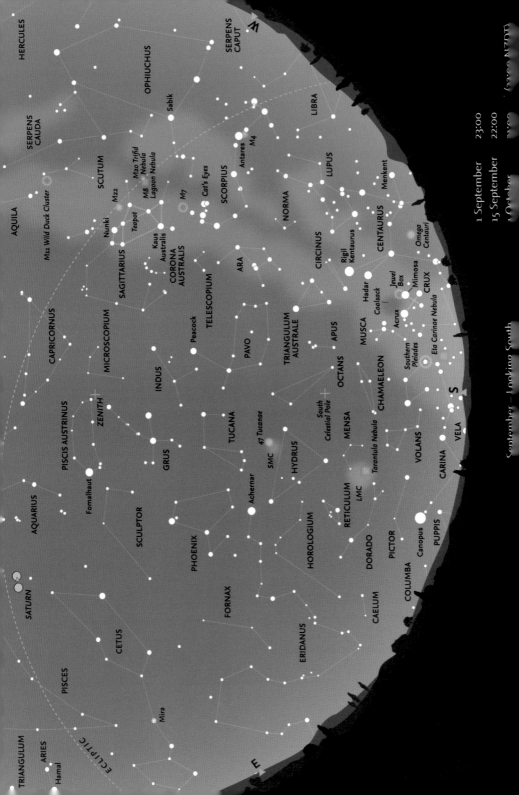

September – Looking South

1 September	23:00
15 September	22:00

September – Looking South

The southern vernal (spring) equinox occurs on 22 September, when the Sun moves south of the equator in *Virgo*.

The smallest of the constellations, **Crux**, is now very low, but **Canopus** (α Carinae), the second brightest star in the sky, has now become visible once more. **Dorado** and the **Large Magellanic Cloud** (LMC) are higher and easier to see, as are the faint constellations of **Pictor**, **Caelum** and **Reticulum**. The **Small Magellanic Cloud** (SMC) and **47 Tucanae** are now nearly halfway between the horizon and the zenith. Although becoming low, the whole of **Centaurus** and **Lupus** remain visible in the southwest. **Scorpius**, with reddish **Antares**, is beginning to descend in the west, but **Sagittarius** is clearly seen high in the sky. The constellation of **Piscis Austrinus**, with **Fomalhaut** (α Piscis Austrini), is at the zenith. Below it are the constellations of **Grus** and **Pavo**, with **Achernar** (α Eridani) and almost the whole of the long constellation of **Eridanus**.

Meteors

There are no major meteor showers active in September. A few meteors may be seen from the **α-Aurigid** shower, active from late August, with a maximum on 31 August. The **Southern Taurid** shower begins this month (on 10 September) and, although rates are low (less than 6 per hour), often produces very bright fireballs. This is a very long shower, lasting until about 20 November. September is notable for a considerable increase in the number of sporadic meteors, which are, of course, completely unpredictable in location, direction and magnitude.

The 'Teapot' of Sagittarius at the top of the image with the curl of stars forming Corona Australis in the lower left, and the 'Cat's Eyes' (Shaula and Lesath) in the 'sting' of Scorpius near the right edge of the image. The bright open cluster M7 (in Scorpius) is prominent between Sagittarius and the Cat's Eyes (north is up).

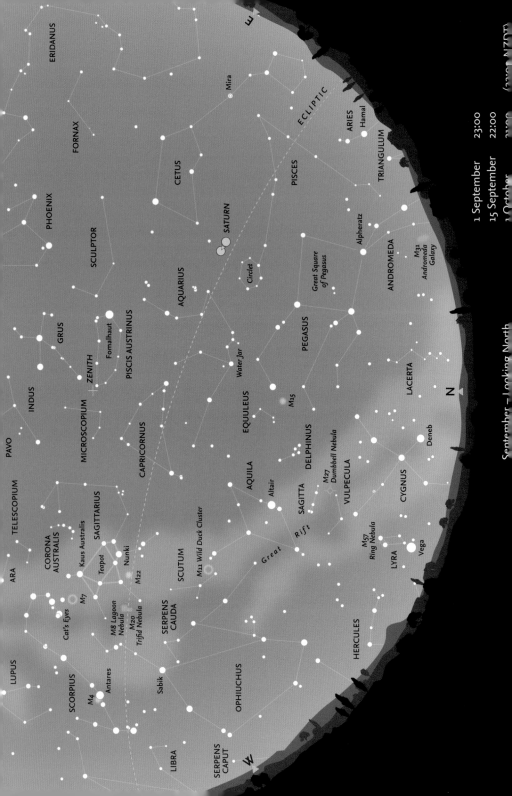

September – Looking North

1 September   23:00
15 September   22:00
1 October   21:00 NZDT

September – Looking North

The Great Square of **Pegasus** is nearing the meridian from the east and, on the west, the three bright stars of the (northern) Summer Triangle, **Deneb** (α Cygni), **Altair** (α Aquilae) and **Vega** (α Lyrae) are still clearly seen, although Vega is low on the horizon, as is **M31** in **Andromeda**. **Delphinus** is well-placed, as are the fainter constellations of **Sagitta** and **Scutum** along the Milky Way. Giant **Ophiuchus** is beginning to descend towards the western horizon and much of **Libra** has disappeared. The three zodiacal constellations of **Pisces**, **Aquarius** and **Capricornus** are readily visible. Somewhat higher, Piscis Austrinus is at the zenith, with Fomalhaut (α Piscis Austrini) prominent. **Scorpius** and **Sagittarius** are now on the western side of the sky.

The constellations of Pisces and Aries. The 'Circlet' of Pisces is top left and the three main stars of Aries, bottom right. In 2025, Uranus will lie in Aries until 2 March (south is up).

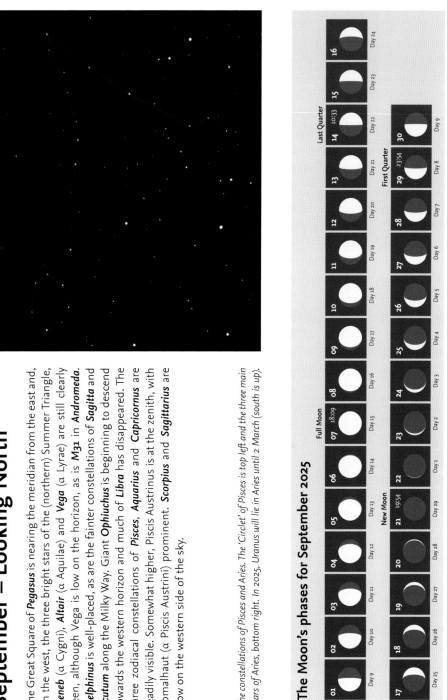

The Moon's phases for September 2025

01 Day 9	02 Day 10	03 Day 11	04 Day 12	05 Day 13	06 Day 14	**Full Moon** 07 18:09 Day 15	08 Day 16
09 Day 17	10 Day 18	11 Day 19	12 Day 20	13 Day 21	**Last Quarter** 14 10:33 Day 22	15 Day 23	16 Day 24
17 Day 25	18 Day 26	19 Day 27	20 Day 28	**New Moon** 21 19:54 Day 29	22 Day 1	23 Day 2	24 Day 3
25 Day 4	26 Day 5	27 Day 6	28 Day 7	**First Quarter** 29 23:54 Day 8	30 Day 9		

September – Moon and Planets

The Moon

A total lunar eclipse on 7 September will be visible from Asia, Russia, Africa, Europe and Australia. The following day the Moon will be 4°N of **Saturn** (mag. 0.6), best seen shortly after midnight. On 14 September the Moon will be Last Quarter; two days later the Moon will pass 4.6°N of **Jupiter** (mag. -2.0) and 2.4°S of **Pollux**. On 19 September **Venus** (mag. -3.9) is occulted by the waning crescent Moon, visible from Africa, western Russia, Canada, Asia and Europe. The Moon will be 1.3°N of **Regulus** in the dawn sky of the same day. Two days after the New Moon on 23 September, the Moon and **Spica** are 1.1° apart, and the following day the crescent Moon and **Mars** (mag. 1.6) will be 3.9° apart. On 27 September the Moon will lie 0.6°S of **Antares** in the early evening sky. The First Quarter Moon will occur on 29 September.

The Planets

Mercury (mag. -1.4 to -0.6) sets shortly after the sunset. **Venus** (mag. -3.9) is visible in the morning sky and moves from **Cancer** to **Leo** on 9 September. **Mars** (mag. 1.6) is best seen just after sunset. **Jupiter** (mag. -2.0 to -2.1) in **Gemini** is visible after midnight. **Saturn** (mag. 0.7 to 0.6) is at opposition on 21 September in **Pisces**. Its rings will appear close to edge-on. **Uranus** (mag. 5.7) enters retrograde motion in **Taurus** on 6 September. **Neptune** (mag. 7.7) will reach opposition on 23 September in **Pisces**.

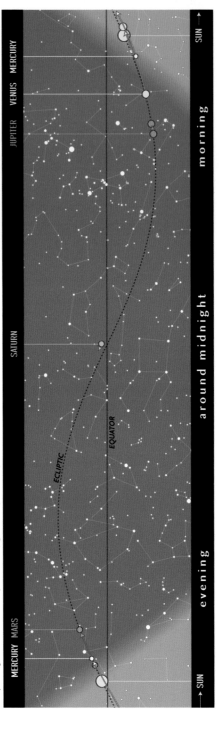

The path of the Sun and the planets along the ecliptic in September.

Calendar for September

07	18:09	FULL MOON
07	18:12	Total Lunar Eclipse; mag.=1.362
08	20:09	Saturn (mag. 0.6) 4.0°S of Moon
10	12:09	Moon at perigee: 364,781 km
12	21:48	Pleiades 1.0°S of Moon
13	03:28	Mars (mag. 1.6) 2.0°N of Spica
14	10:33	LAST QUARTER MOON
16	11:06	Jupiter (mag. –2.0) 4.6°S of Moon
16	17:58	Pollux 2.4°N of Moon
19	08:57	Venus (mag. –3.9) 0.4°N of Regulus
19	11:11	Regulus 1.3°S of Moon
19	11:46	Venus (mag. –3.9) 0.8°S of Moon
21	05:00	Saturn at opposition
21	19:42	Partial Solar Eclipse; mag.=0.855
21	19:54	NEW MOON
22	18:20	Autumnal Equinox
23	11:00	Neptune at opposition (mag. 7.7)
23	21:31	Spica 1.1°N of Moon
24	14:50	Mars (mag. 1.6) 3.9°N of Moon
26	09:46	Moon at apogee: 405,552 km
27	17:34	Antares 0.6°N of Moon
28		Daylight Saving Time begins (New Zealand)
29	23:54	FIRST QUARTER MOON

Evening 21:00

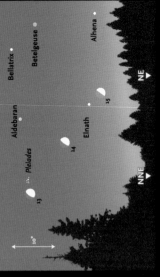

8–9 September • The Moon passes Saturn (mag. 0.6). Diphda (β Cet) is due east.

Early morning 3:00

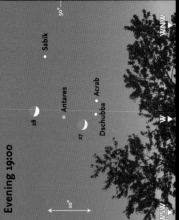

13–15 September • The Moon passes the Pleiades and Elnath. Aldebaran and some of the bright stars of the constellation of Orion are higher in the northeast.

Morning 5:00

16–17 September • The Moon passes between Jupiter (mag. –2.1) and the Twin Stars (Castor and Pollux). Alhena and Procyon are both higher in the northeast.

Evening 19:00

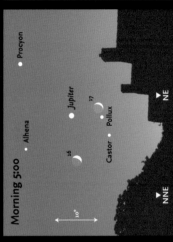

27–28 September • High in the west, the Moon passes Antares. The name means 'Rival of Mars' because the red colour resembles the colour of Mars.

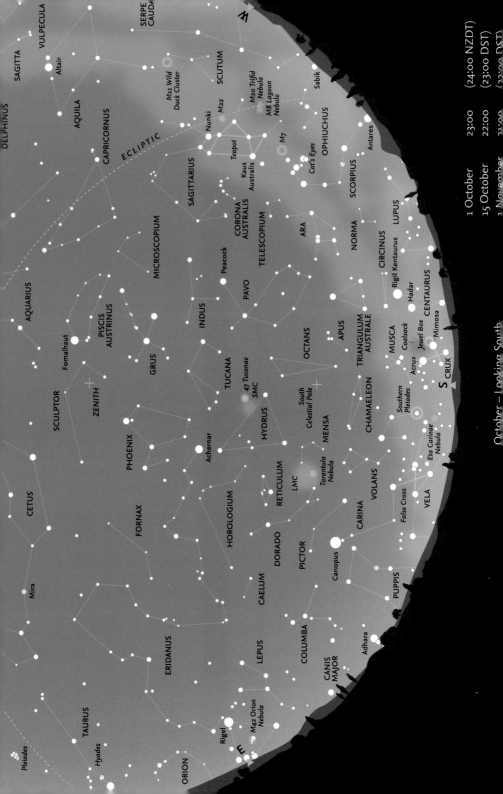

October – Looking South

1 October   23:00   (24:00 NZDT)
15 October   22:00   (23:00 DST)
1 November   [22:00]   (22:00 DST)

October – Looking South

Crux is now extremely low, only just visible above the horizon. The *False Cross* has reappeared, also very low, farther towards the east on the opposite side of the meridian. *Canopus* (α Carinae) is now much higher as is the *Large Magellanic Cloud* (LMC). The *Small Magellanic Cloud* (SMC) and the globular cluster *47 Tucanae* are on the southern meridian, halfway to the zenith, as is *Achernar* (α Eridani). The whole of the long constellation of *Eridanus* is now visible together with *Rigel* in *Orion*. Still higher are *Phoenix*, *Grus* and *Piscis Austrinus*. *Pavo* has begun to descend in the southwest. *Ophiuchus* has now slipped below the horizon as has much of *Scorpius*, only the 'tail' of which remains visible. *Sagittarius* is getting lower, but remains visible, as do the zodiacal constellations of *Capricornus* and *Aquarius*.

Meteors

The *Orionids* are the major, fairly reliable meteor shower active in October. Like the May *η-Aquariid* shower, the Orionids are associated with Comet 1P/Halley. During this second pass through the stream of particles from the comet, slightly fewer meteors are seen than in May. In both showers the meteors are very fast and many leave persistent trains. Although the Orionid maximum is nominally 22 October, in fact there is a very broad maximum, lasting about a week from 20 to 27 October, with hourly rates around 15. Occasionally, rates are higher (50–70 per hour). In 2025, there is a New Moon on the night of the peak, so conditions are favourable. The faint shower of the *Southern Taurids* (often with bright fireballs) peaks on 10 October. The Southern Taurid maximum occurs when the Moon is waning gibbous, conditions are unfavourable from a few hours before midnight when the Moon rises. Towards the end of the month (around 20 October), another shower (the *Northern Taurids*) begins to show activity, which peaks early in November. The radiants for both Taurid showers are close together.

The combined position is shown on the 'Looking North' chart. The parent comet for both Taurid showers is Comet 2P/Encke.

The Small Magellanic Cloud, one of the satellite galaxies of the Milky Way. Near the right edge of this image is 47 Tucanae (NGC 104), which is the second brightest globular cluster in the sky, after Omega Centauri (see page 53) (north is up).

October – Looking North

1 October   23:00   (24:00 NZDT)
15 October   22:00   (23:00 DST)
1 November   (22:00 DST)

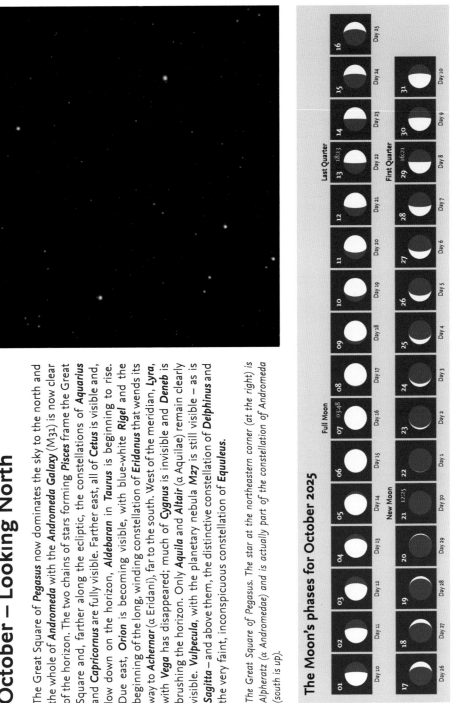

October – Looking North

The Great Square of **Pegasus** now dominates the sky to the north and the whole of **Andromeda** with the **Andromeda Galaxy** (M31) is now clear of the horizon. The two chains of stars forming **Pisces** frame the Great Square and, farther along the ecliptic, the constellations of **Aquarius** and **Capricornus** are fully visible. Farther east, all of **Cetus** is visible and, low down on the horizon, **Aldebaran** in **Taurus** is beginning to rise. Due east, **Orion** is becoming visible, with blue-white **Rigel** and the beginning of the long, winding constellation of **Eridanus** that wends its way to **Achernar** (α Eridani), far to the south. West of the meridian, **Lyra**, with **Vega** has disappeared; much of **Cygnus** is invisible and **Deneb** is brushing the horizon. Only **Aquila** and **Altair** (α Aquilae) remain clearly visible. **Vulpecula**, with the planetary nebula **M27** is still visible – as is **Sagitta** – and above them, the distinctive constellation of **Delphinus** and the very faint, inconspicuous constellation of **Equuleus**.

The Great Square of Pegasus. The star at the northeastern corner (at the right) is Alpheratz (α Andromedae) and is actually part of the constellation of Andromeda (south is up).

The Moon's phases for October 2025

01	02	03	04	05	06	07 03:48	08	09	10	11	12	13 18:13	14	15	16
Day 10	Day 11	Day 12	Day 13	Day 14	Day 15	Day 16	Day 17	Day 18	Day 19	Day 20	Day 21	Day 22	Day 23	Day 24	Day 25

Full Moon — 07 03:48
Last Quarter — 13 18:13

17	18	19	20	21 12:25	22	23	24	25	26	27	28	29 16:21	30	31
Day 26	Day 27	Day 28	Day 29	Day 30	Day 1	Day 2	Day 3	Day 4	Day 5	Day 6	Day 7	Day 8	Day 9	Day 10

New Moon — 21 12:25
First Quarter — 29 16:21

October – Moon and Planets

The Moon

On 6 October the Moon and **Saturn** (mag. 0.6) will be 3.8° apart, a day later the Moon will be full. On the 13th the Moon will be Last Quarter and will lie just over 4°N of **Jupiter** (mag. -2.2) and 2.5°S of **Pollux**. Three days later the Moon is 1.3°N of **Regulus**. On 19 October the waning crescent Moon will pass 3.7°S of **Venus** (mag. -3.9), visible in the dawn sky. After the New Moon on 21 October, the thin crescent Moon will be 4.5°S of **Mars** (mag. 1.4) and two days later it will appear 2.3°S of **Mercury**. On 25 October the Moon moves less than a degree from **Antares** and on 29 October the Moon will be First Quarter.

The planets

Mercury is at greatest eastern elongation on 29 October, at mag. -0.1. In **Libra**, Mercury passes just over 2°S of Mars on 21 October, during the day. **Venus** (mag. -3.9) moves from **Leo** to **Virgo** on 8 October, visible shortly before sunrise. **Mars** (mag. 1.6 to 1.4) sets shortly after the sunset in **Libra**. **Jupiter** (mag. -2.1 to -2.3) is bright in the dawn sky lying in **Gemini**. **Saturn** (mag. 0.6 to 0.8) is visible in **Aquarius** throughout the night. **Uranus** (mag. 5.7 to 5.6) is in **Taurus** and **Neptune** (mag. 7.7) is in **Pisces**. Minor planet (1) **Ceres** (mag. 7.6) is at opposition on 3 October.

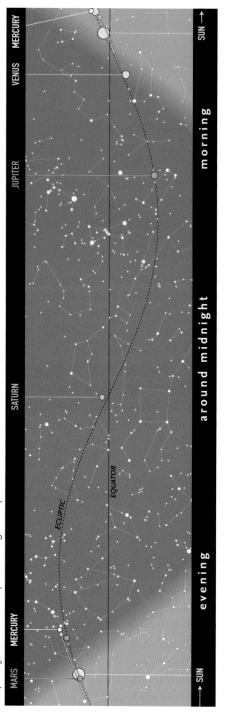

The path of the Sun and the planets along the ecliptic in October.

Calendar for October

02	11:00	Venus at perihelion
02–07		Orionid meteor shower
03	03:23	Dwarf planet (1) Ceres at opposition (mag. 7.6)
06	02:46	Saturn (mag. 0.6) 3.8°S of Moon
07	03:48	FULL MOON
08	12:36	Moon at perigee: 359,819 km
10	05:20	Pleiades 0.9°S of Moon
10		Southern Taurid Meteor Shower maximum
13	18:13	LAST QUARTER MOON
13	22:31	Jupiter (mag. -2.2) 4.3°S of Moon
13	23:31	Pollux 2.5°N of Moon
16	16:56	Regulus 1.3°S of Moon
19	20:00	Mercury (mag. -0.2) 2.0°S of Mars
19	21:37	Venus (mag. -3.9) 3.7°N of Moon
21	12:25	NEW MOON
22		Orionid Meteor Shower maximum
23	16:15	Mercury (mag. -0.2) 2.3°N of Moon
23	23:31	Moon at apogee: 406,445 km
25	00:15	Antares 0.5°N of Moon
29	16:21	FIRST QUARTER MOON
29	22:00	Mercury at greatest elong: 23.9°E

Early morning 4:00 (DST)

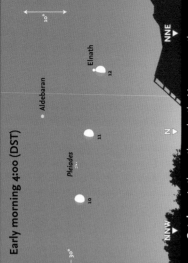

6–7 October • The Moon is almost Full, when it passes Saturn (mag. 0.6).

Early morning 4:00 (DST)

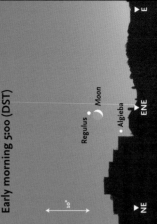

10–12 October • In the northern sky, the Moon passes the Pleiades and Elnath. Aldebaran is about ten degrees higher.

Early morning 4:00 (DST)

14 October • The Last Quarter Moon is near Jupiter (mag. -2.3) and the Twin Stars (Castor and Pollux). Alhena and Procyon are both higher in the northeast.

Early morning 5:00 (DST)

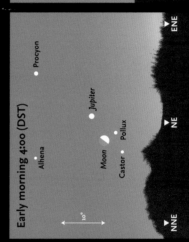

17 October • In the early morning the crescent Moon passes between Regulus and Algieba (γ Leo), which is too close to the horizon to be easily seen.

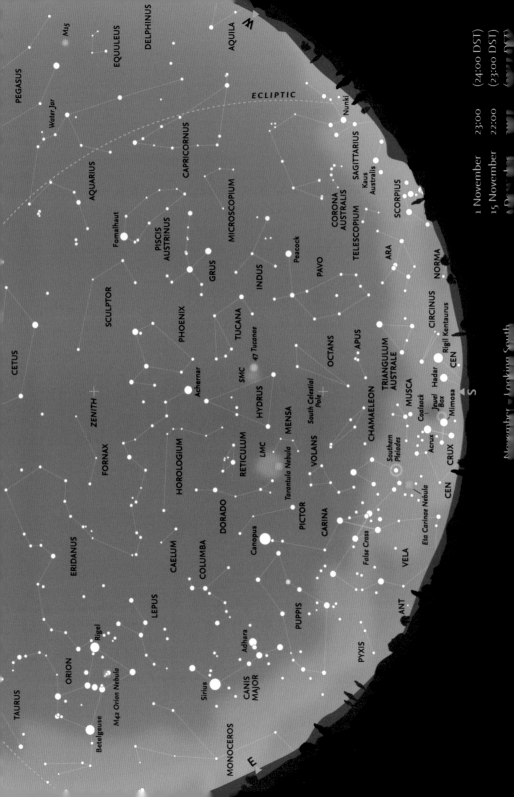

November – Looking South

1 November	23:00	(24:00 DST)
15 November	22:00	(23:00 DST)

November – Looking South

The constellation of Canis Major, with Sirius, the brightest star in the sky (mag. -1.4) in the upper half of the image. Below Sirius is the open cluster M41 (north is up).

Crux and the two brightest stars of **Centaurus, Rigil Kentaurus** (α Centauri) and **Hadar** (β Centauri), are extremely low on the southern horizon. The **False Cross** on the **Carina/Vela** border is now higher, and the constellation of **Puppis** as well as **Canopus** (α Carinae), and both the **Large Magellanic Cloud** (LMC) and the **Small Magellanic Cloud** (SMC), are clearly visible. **Achernar** (α Eridani) is halfway between the South Celestial Pole and the zenith. The whole of **Eridanus**, which starts near **Rigel** in **Orion**, is now clearly seen as it winds its way to Achernar. **Pavo** is becoming lower in the southwest and, in the west, most of **Sagittarius** is below the horizon, with **Capricornus** descending behind it. **Corona Australis** is still just visible. In the east, **Canis Major** is now clearly seen, together with the small constellations of **Columba** and **Lepus** above it.

Meteors

Two meteor showers begin in September or October but continue into November. The **Orionids** (see page 89), one of the streams associated with Comet 1/P Halley, continue until about 7 November. The **Southern Taurids** (see page 83) begin on 10 September and continue until 20 November. Because of the location of the radiant, the **Leonid** shower is best seen from the northern hemisphere, but southern observers may see some rising from the horizon. They have a short period of activity (6–30 November), with maximum on 17 November. This shower is associated with Comet 55P/Tempel-Tuttle and has shown extraordinary activity on various occasions with many thousands of meteors per hour. The rate in 2025 is likely to be about 15 per hour. These meteors are the fastest shower meteors recorded (about 70 km per second) and often leave persistent trains. The shower is very rich in faint meteors.

There is a minor southern meteor shower that begins activity in late November (nominally the 28th). This is the **Phoenicids**, but little is known of the shower, partly because the parent comet is believed to be the disintegrated comet D/1819 W1 (Blanpain). With no accurate knowledge of the location of the remnants of the comet, predicting the possible rate becomes little more than guesswork, but the rate is variable and may rapidly increase (as might be expected) if the orbit is nearby. Bright meteors tend to be quite frequent and the meteors are fairly slow. The radiant is located within **Phoenix**, not far from the border with **Eridanus** and the bright star **Achernar** (α Eridani).

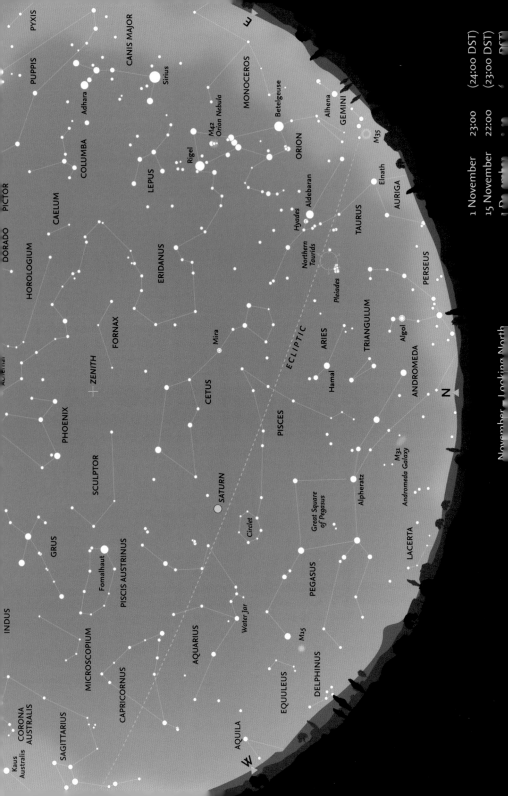

November – Looking North

November – Looking North

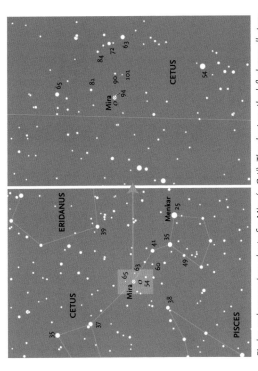

Andromeda is now due north, and the **Andromeda Galaxy** (M31) has risen sufficiently to be clearly seen. The constellation of **Triangulum** lies between Andromeda and the zodiacal constellation of **Aries**. Much of **Perseus** (including the variable star, **Algol**) is now above the horizon, together with part of **Auriga**. The whole of **Pegasus**, with the **Great Square**, is clearly seen and above it the two lines of stars forming **Pisces** and the distinctive asterism of the **Circlet**. Higher still is the constellation of **Cetus** with the famous variable star, **Mira**. The whole of **Taurus** with **Aldebaran** (α Tauri), and the **Pleiades** and the **Hyades** clusters are visible in the southeast. **Orion** has fully risen in the east with **Lepus** above it. The long, winding constellation of **Eridanus** begins near **Rigel** in Orion. The faint constellation of **Monoceros** lies to the east of Orion and straddles the Milky Way. **Canis Major** and brilliant **Sirius** are even farther round towards the east. In the west, the zodiacal constellations of **Aquarius** (with its distinctive asterism of the 'Water Jar') and **Capricornus** are clearly seen, with **Piscis Austrinus** and bright **Fomalhaut** (α Piscis Austrini) higher in the sky. The faint constellation of **Sculptor** lies between Piscis Austrinus and the zenith.

Finder and comparison charts for Mira (o Ceti). The chart on the left shows all stars brighter than magnitude 6.5. The chart on the right shows stars down to magnitude 10.0. The comparison star magnitudes are shown without the decimal point (south is up).

The Moon's phases for November 2025

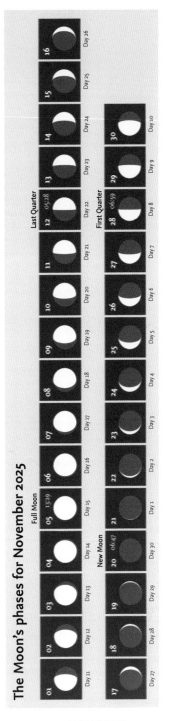

November – Moon and Planets

The Moon

On 2 November the waxing gibbous Moon will lie 3.7°N of **Saturn** (mag. 0.8). On 5 November the Full Moon will be at its closest point to the Earth – its perigee – a distance of 356,833 km. On 10 November the bright waning gibbous Moon will pass within 4°N of **Jupiter** (mag. -2.4), best seen after midnight. **Pollux** will be less than 3° away. Two days later the Moon will be Last Quarter and around 1°S of **Regulus.** On 17 November **Spica** will be 1.2°N of the crescent Moon, both setting before the Sun. On 20 November the New Moon will be at its farthest point – apogee, 406,693 km from Earth. The following day the thin crescent Moon will be 4.5°S of **Mars** (mag. 1.4) during the day. On 29 November, a day after First Quarter, the Moon will appear 3.7°N of **Saturn** (mag. 0.9).

The Planets

Mercury (mag. -0.7 to 0.4) is above the horizon during the day, it passes just over a degree from **Mars** on 13 November during this time. **Venus** (mag. -3.9) appears in the dawn sky, setting just before the Sun. **Mars** (mag. 1.5 to 1.3) is seen just after sunset in **Libra. Jupiter** (mag. -2.3 to -2.5) climbs above the horizon late in the night and enters retrograde motion on 11 November, visible in **Gemini. Saturn** (mag. 0.8 to 0.9) ends its retrograde motion towards the end of the month in **Aquarius** and starts to move eastward. **Uranus** is at opposition on 21 November (mag. 5.6) in **Taurus** and **Neptune** (mag. 7.7) is in **Pisces.**

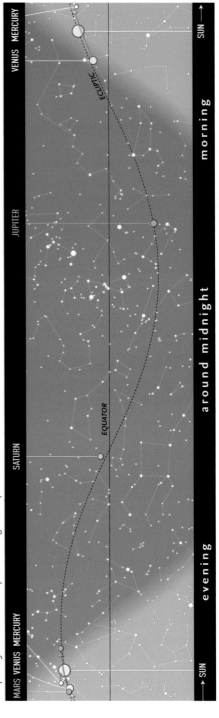

The path of the Sun and the planets along the ecliptic in November.

Calendar for November

02	01:02	Venus (mag. -3.9) 3.3°N of Spica
02	10:46	Saturn (mag. 0.8) 3.7°S of Moon
05	13:19	FULL MOON
05	22:29	Moon at perigee: 356,833 km
06	15:26	Pleiades 0.8°S of Moon
09	02:41	Mercury (mag. 0.3) 2.6°N of Antares
10	06:40	Pollux 2.7°N of Moon
10	07:56	Jupiter (mag. -2.4) 4.0°S of Moon
12	05:28	LAST QUARTER MOON
12		Northern Taurid Meteor Shower maximum
12	22:51	Regulus 1.1°S of Moon
13	04:00	Mercury (mag. 1.1) 1.2°S of Mars (mag. 1.5)
17	10:11	Spica 1.2°N of Moon
17		Leonid Meteor Shower maximum
20	02:48	Moon at apogee: 406,693 km
20	06:47	NEW MOON
21	13:00	Uranus at opposition (mag. 5.6)
23	11:00	Mercury at perihelion
28	06:59	FIRST QUARTER MOON

After midnight 2:00 (DST)

2–3 November • The Moon passes Saturn (mag. 0.8) in the western sky.

After midnight 2:00 (DST)

10–11 November • Low in the northeast, the Moon passes between Castor, Pollux, and Jupiter (mag. -2.4). Alhena and Procyon are both fifteen degrees higher.

Early morning 4:00 (DST)

13–14 November • The crescent Moon passes between Regulus and Algieba.

Early morning 5:00 (DST)

17–18 November • The Moon passes Spica, low in the eastern sky.

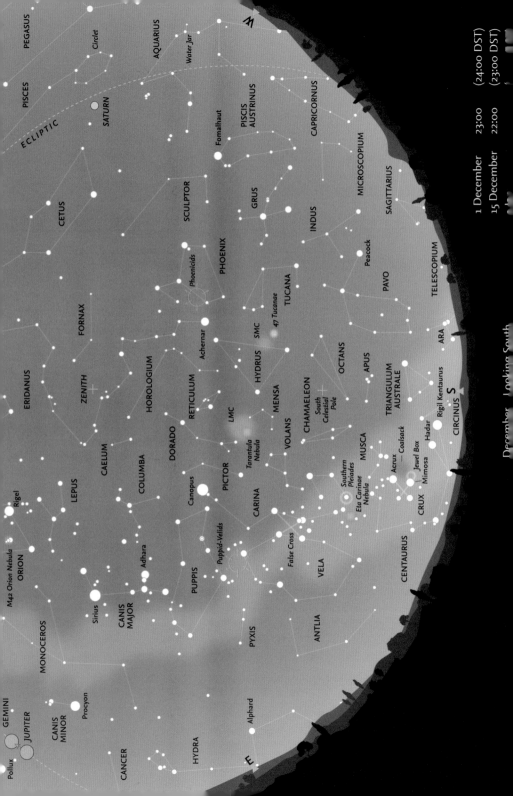

December, Looking South

December – Looking South

The Large Magellanic Cloud seen over the European Southern Observatory at Paranal in Chile. Brilliant Canopus (α Carinae), the second brightest star in the southern hemisphere, is visible through the low cloud to the right.

Crux and the two brightest stars in **Centaurus**, **Hadar** and **Rigil Kentaurus**, are now higher above the horizon. The **Eta Carina Nebula** and the **Southern Pleiades** are now conveniently placed for observation. Above them, the **False Cross** is clearly seen, with the whole of **Vela** and below it the inconspicuous constelation of **Antlia**. **Carina** with **Canopus** (α Carinae) and **Puppis** are roughly halfway between the horizon and the zenith. **Sirius** and **Canis Major** are high in the east. In the west, **Achernar** (α Eridani) and **Phoenix** are about the same altitude as Canopus. The faint constellations of **Pictor**, **Dorado**, **Reticulum** and **Horologium** lie between them. **Pavo** with **Peacock** (α Pavonis) is becoming low, as are the constellations of **Indus**, **Grus** and **Piscis Austrinus**. Higher still are the constellations of **Sculptor** and, near the zenith, **Fornax**. **Capricornus** is largely invisible, but most of **Aquarius** may still be seen.

Meteors

The **Phoenicid** shower continues into December, reaching its weak maximum on 2 December. The **Puppid-Velid** shower's radiant is on the border between the two constellations. The shower begins on 1 December, lasting until 15 December, with maximum on 7 December. It is a weak shower with a maximum hourly rate of about 10 meteors, but bright meteors are often seen. One of the most dependable showers of the year is the **Geminids**, active from 4 to 20 December, with maximum in 2025 on 14 December. The rate is often 60–70 per hour, and may rise even higher. The Moon is waning crescent; conditions will be moderately favourable once the Moon rises after midnight.

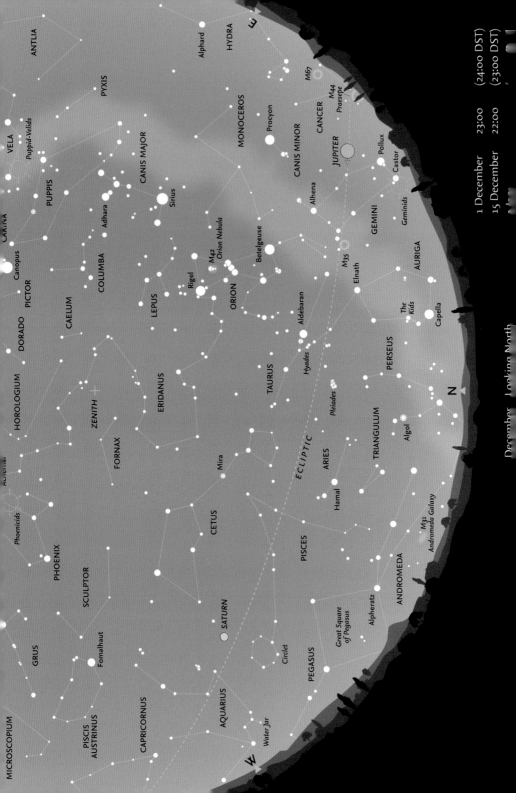

December Looking North

1 December   23:00   (24:00 DST)
15 December   22:00   (23:00 DST)

December – Looking North

Most of *Perseus* may be seen due north with, above it, the *Pleiades* cluster. Farther west, **Andromeda** is very low and the southern stars of *Pegasus* have been lost below the horizon. Above Pegasus lies the zodiacal constellation of *Pisces* and, still higher, **Cetus** and the faint constellatons of **Sculptor** and *Fornax*. The famous variable of *Mira* in Cetus (see charts on page 97) is ideally placed for observation. Towards the east, the whole of the constellations of both **Auriga** and **Gemini** are visible, although **Capella** (α Aurigae) and **Castor** and **Pollux** (α and β Gemini) are low on the horizon. **Taurus, Orion** and **Canis Minor** are readily visible, together with the faint constellation of **Monoceros.** *Eridanus* wanders from its start near **Rigel** (β Orionis) towards **Achernar** (α Eridani), beyond the zenith.

The constellation of Taurus contains two contrasting open clusters: the compact Pleiades, with its striking blue-white stars, and the more scattered, 'V'-shaped Hyades, which are much closer to us. Orange Aldebaran (α Tauri) is not related to the Hyades, but lies between it and the Earth (south is up).

The Moon's phases for December 2025

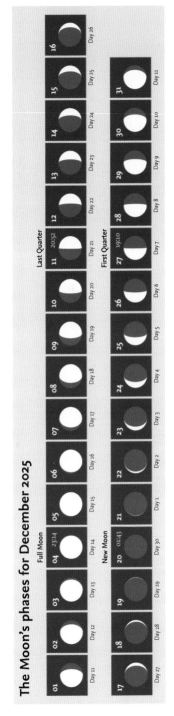

December – Moon and Planets

The Moon

On 4 December the Moon is Full, three days later it will be 3.7°N of **Jupiter** (mag. -2.6) and less than 3°S of **Pollux**. On 10 December the waning gibbous Moon is less than a degree from **Regulus**, and the following day the Moon is Last Quarter. On 14 December the crescent Moon will be 1.4°S of **Spica** and on 20 December it will be a New Moon. On 27 December the First Quarter Moon will lie 4°N of **Saturn** (mag. 1.0), visible in the early evening.

The planets

Mercury increases in apparent brightness as the month progresses (mag. 0.07 to -0.6), on 8 December it will be at greatest western elongation. **Venus** (mag. -3.9) moves into **Virgo** early in the month, visible at dawn. **Mars** (mag. 1.3 to 1.1) moves from **Ophiuchius** to **Sagittarius** around the middle of the month, setting shortly after the Sun. **Jupiter** (mag. -2.5 to -2.7) stays in **Gemini**. **Saturn** (mag. 0.9 to 1.0) continues to move eastward in **Aquarius**, setting earlier over the course of the month. **Uranus** (mag. 5.6) is in **Taurus** and **Neptune** (mag. 7.7 to 7.8) ends its westward retrograde motion in **Pisces** on 10 December.

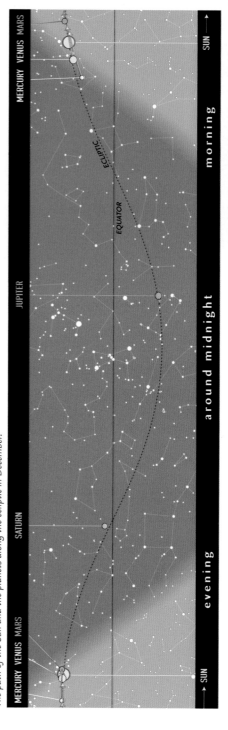

The path of the Sun and the planets along the ecliptic in December.

After midnight 0:30 (DST)

10° NNE   NE ►   ENE ►

Alhena • Procyon

Jupiter

Moon • Pollux

Castor •

8 December • *Low in the northeast, the Moon is in the company of Jupiter, Pollux and Castor. Both Alhena and Procyon are about fifteen degrees higher.*

After midnight 2:00 (DST)

10°   ◄ NE   ENE ►

Regulus
11
10
Algieba

10–11 December • *The waning gibbous Moon passes between Regulus and Algieba, in Leo.*

Early morning 3:30 (DST)

10°   ◄ E   ESE ►

Spica • Moon

15 December • *In the early morning the Moon passes Spica, almost due east.*

Evening 23:00 (DST)

10°   ◄ W   WNW ►

Saturn •
27
26

26–27 December • *Low in the western sky, the First Quarter Moon passes Saturn (mag. 1.0).*

Phoenicid Meteor Shower maximum

Pleiades 0.8°S of Moon

Moon at perigee: 356,962 km

FULL MOON

Jupiter (mag. –2.6) 3.7°S of Moon

Pollux 2.9°N of Moon

Mercury at greatest elong: 20.7°W

Puppid-Velid Meteor Shower maximum

Regulus 0.8°S of Moon

LAST QUARTER MOON

Geminid Meteor Shower maximum

Spica 1.4°N of Moon

Moon at apogee: 406,324 km

Antares 0.4°N of Moon

NEW MOON

Summer Solstice

Saturn (mag. 1.0) 4.0°S of Moon

FIRST QUARTER MOON

Pleiades 0.9°S of Moon

Dark Sky Sites

International Dark-Sky Association Sites

The *International Dark-Sky Association* (IDA) recognizes various categories of sites that offer areas where the sky is dark at night, free from light pollution and particularly suitable for astronomical observing. Although a number of sites are under consideration, the majority of confirmed sites are in North America. There are more than 90 in the United States and Canada. There are 20 sites in Great Britain and Ireland, seven in Europe, and one each in Israel, Japan and South Korea. There are at least 13 sites in the southern hemisphere and the list is growing.

Details of the IDA are at: https://www.darksky.org/. Information on the various categories and individual sites are at: https://www.darksky.org/our-work/conservation/idsp/. New sites are continually being added, so check the IDA website for details.

Many of these sites have major observatories or other facilities available for public observing (often at specific dates or times).

This montage of images from NASA's Earth-orbiting satellites was obtained in 2016. Since then the situation in the northern hemisphere has become even worse for observers. Markers show the location of the 13 Dark-Sky sites in the southern hemisphere.

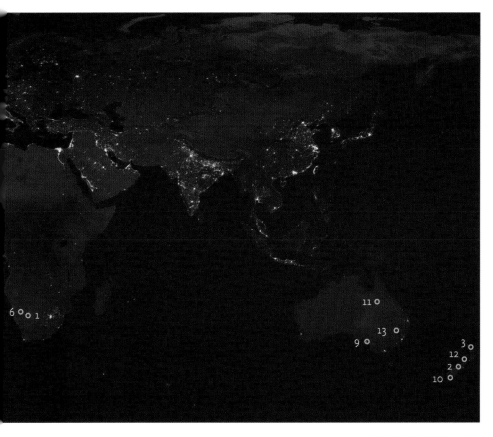

Glossary and Tables

aphelion	The point on an orbit that is farthest from the Sun.
apogee	The point on its orbit at which the Moon is farthest from the Earth.
appulse	The apparently close approach of two celestial objects; two planets, or a planet and star.
astronomical unit	(AU) The mean distance of the Earth from the Sun, 149,597,870 km.
celestial equator	The great circle on the celestial sphere that is in the same plane as the Earth's equator.
celestial sphere	The apparent sphere surrounding the Earth on which all celestial bodies (stars, planets, etc.) seem to be located.
conjunction	The point in time when two celestial objects have the same celestial longitude. In the case of the Sun and a planet, superior conjunction occurs when the planet lies on the far side of the Sun (as seen from Earth). For Mercury and Venus, inferior conjunction occurs when they pass between the Sun and the Earth.
direct motion	Motion from west to east on the sky.
ecliptic	The apparent path of the Sun across the sky throughout the year. Also: the plane of the Earth's orbit in space.
elongation	The point at which an inferior planet has the greatest angular distance from the Sun, as seen from Earth.
equinox	The two points during the year when night and day have equal duration. Also: the points on the sky at which the ecliptic intersects the celestial equator. The vernal (spring) equinox is of particular importance in astronomy.
gibbous	The stage in the sequence of phases at which the illumination of a body lies between half and full. In the case of the Moon, the term is applied to phases between First Quarter and Full, and between Full and Last Quarter.
inferior planet	Either of the planets Mercury or Venus, which have orbits inside that of the Earth.
magnitude	The brightness of a star, planet or other celestial body. It is a logarithmic scale, where larger numbers indicate fainter brightness. A difference of 5 in magnitude indicates a difference of 100 in actual brightness, thus a first-magnitude star is 100 times as bright as one of sixth magnitude.
meridian	The great circle passing through the North and South Poles of a body and the observer's position; or the corresponding great circle on the celestial sphere that passes through the North and South Celestial Poles and also through the observer's zenith.
nadir	The point on the celestial sphere directly beneath the observer's feet, opposite the zenith.
occultation	The disappearance of one celestial body behind another, such as when stars or planets are hidden behind the Moon.
opposition	The point on a superior planet's orbit at which it is directly opposite the Sun in the sky.
perigee	The point on its orbit at which the Moon is closest to the Earth.
perihelion	The point on an orbit that is closest to the Sun.
retrograde motion	Motion from east to west on the sky.
superior planet	A planet that has an orbit outside that of the Earth.
vernal equinox	The point at which the Sun, in its apparent motion along the ecliptic, crosses the celestial equator from south to north. Also known as the First Point of Aries.
zenith	The point directly above the observer's head.
zodiac	A band, stretching 8° on either side of the ecliptic, within which the Moon and planets appear to move. It consists of twelve equal areas, originally named after the constellation that once lay within it.

The Constellations

There are 88 constellations covering the whole of the celestial sphere, but 4 of these in the northern hemisphere (Camelopardalis, Cassiopeia, Cepheus and Ursa Minor) can never be seen (even in part) from a latitude of 35°S, so are omitted from this table. The names themselves are expressed in Latin, and the names of stars are frequently given by Greek letters (see next page) followed by the genitive of the constellation name. The genitives and English names of the various constellations are included.

Name	Genitive	Abbr.	English name
Andromeda	Andromedae	And	Andromeda
Antlia	Antliae	Ant	Air Pump
Apus	Apodis	Aps	Bird of Paradise
Aquarius	Aquarii	Aqr	Water Bearer
Aquila	Aquilae	Aql	Eagle
Ara	Arae	Ara	Altar
Aries	Arietis	Ari	Ram
Auriga	Aurigae	Aur	Charioteer
Boötes	Boötis	Boo	Herdsman
Caelum	Caeli	Cae	Burin
Cancer	Cancri	Cnc	Crab
Canes Venatici	Canum Venaticorum	CVn	Hunting Dogs
Canis Major	Canis Majoris	CMa	Big Dog
Canis Minor	Canis Minoris	CMi	Little Dog
Capricornus	Capricorni	Cap	Sea Goat
Carina	Carinae	Car	Keel
Centaurus	Centauri	Cen	Centaur
Cetus	Ceti	Cet	Whale
Chamaeleon	Chamaeleontis	Cha	Chameleon
Circinus	Circini	Cir	Compasses
Columba	Columbae	Col	Dove
Coma Berenices	Comae Berenices	Com	Berenice's Hair
Corona Australis	Coronae Australis	CrA	Southern Crown
Corona Borealis	Coronae Borealis	CrB	Northern Crown
Corvus	Corvi	Crv	Crow
Crater	Crateris	Crt	Cup
Crux	Crucis	Cru	Southern Cross
Cygnus	Cygni	Cyg	Swan
Delphinus	Delphini	Del	Dolphin
Dorado	Doradus	Dor	Dorado
Draco	Draconis	Dra	Dragon
Equuleus	Equulei	Equ	Little Horse
Eridanus	Eridani	Eri	River Eridanus
Fornax	Fornacis	For	Furnace
Gemini	Geminorum	Gem	Twins
Grus	Gruis	Gru	Crane
Hercules	Herculis	Her	Hercules
Horologium	Horologii	Hor	Clock
Hydra	Hydrae	Hya	Water Snake
Hydrus	Hydri	Hyi	Lesser Water Snake
Indus	Indi	Ind	Indian
Lacerta	Lacertae	Lac	Lizard

Name	Genitive	Abbr.	English name
Leo	Leonis	Leo	Lion
Leo Minor	Leonis Minoris	LMi	Little Lion
Lepus	Leporis	Lep	Hare
Libra	Librae	Lib	Scales
Lupus	Lupi	Lup	Wolf
Lynx	Lyncis	Lyn	Lynx
Lyra	Lyrae	Lyr	Lyre
Mensa	Mensae	Men	Table Mountain
Microscopium	Microscopii	Mic	Microscope
Monoceros	Monocerotis	Mon	Unicorn
Musca	Muscae	Mus	Fly
Norma	Normae	Nor	Set Square
Octans	Octantis	Oct	Octant
Ophiuchus	Ophiuchi	Oph	Serpent Bearer
Orion	Orionis	Ori	Orion
Pavo	Pavonis	Pav	Peacock
Pegasus	Pegasi	Peg	Pegasus
Perseus	Persei	Per	Perseus
Phoenix	Phoenicis	Phe	Phoenix
Pictor	Pictoris	Pic	Painter's Easel
Pisces	Piscium	Psc	Fishes
Piscis Austrinus	Piscis Austrini	PsA	Southern Fish
Puppis	Puppis	Pup	Stern
Pyxis	Pyxidis	Pyx	Compass
Reticulum	Reticuli	Ret	Net
Sagitta	Sagittae	Sge	Arrow
Sagittarius	Sagittarii	Sgr	Archer
Scorpius	Scorpii	Sco	Scorpion
Sculptor	Sulptoris	Scu	Sculptor
Scutum	Scuti	Sct	Shield
Serpens	Serpentis	Ser	Serpent
Sextans	Sextantis	Sex	Sextant
Taurus	Tauri	Tau	Bull
Telescopium	Telescopii	Tel	Telescope
Triangulum	Trianguli	Tri	Triangle
Triangulum Australe	Trianguli Australis	TrA	Southern Triangle
Tucana	Tucanae	Tuc	Toucan
Ursa Major	Ursae Majoris	UMa	Great Bear
Vela	Velorum	Vel	Sails
Virgo	Virginis	Vir	Virgin
Volans	Volantis	Vol	Flying Fish
Vulpecula	Vulpeculae	Vul	Fox

The Greek Alphabet

α	Alpha	ε	Epsilon	ι	Iota	ν	Nu	ρ	Rho	φ (φ) Phi
β	Beta	ζ	Zeta	κ	Kappa	ξ	Xi	σ (ς)	Sigma	χ Chi
γ	Gamma	η	Eta	λ	Lambda	ο	Omicron	τ	Tau	ψ Psi
δ	Delta	θ (ϑ)	Theta	μ	Mu	π	Pi	υ	Upsilon	ω Omega

Some common asterisms

Belt of Orion	δ, ε and ζ Orionis
Cat's Eyes	λ and υ Scorpii
Circlet	γ, θ, ι, λ and κ Piscium
False Cross	ε and ι Carinae and δ and κ Velorum
Fish Hook	α, β, δ and π Scorpii
Head of Cetus	α, γ, ξ², μ and λ Ceti
Head of Hydra	δ, ε, ζ, η, ρ and σ Hydrae
Job's Coffin	α, β, γ and δ Delphini
Keystone	ε, ζ, η and π Herculis
Kids	ζ and η Aurigae
Milk Dipper	ζ, γ, σ, φ and λ Sagittarii
Pot	= Saucepan
Saucepan	ι, θ, ζ, ε, δ and η Orionis
Sickle	α, η, γ, ζ, μ and ε Leonis
Southern Pointers	α and β Centauri
Square of Pegasus	α, β and γ Pegasi with α Andromedae
Sword of Orion	θ and ι Orionis
Teapot	γ, ε, δ, λ, φ, σ, τ and ζ Sagittarii
Water Jar	γ, η, κ and ζ Aquarii
Y of Aquarius	= Water Jar

Acknowledgements

Denis Buczynski, Portmahomack, Ross-shire – p.32 (Fireball)

Dave Chapman, Dartmouth, Nova Scotia, Canada – p.77 (LMC)

European Southern Observatory – pp.53 (Omega Centauri), 89, 101 (Magellanic Clouds)

Akira Fuji – p.21 (Lunar eclipse)

Bernhard Hubl – pp.35, 43, 47, 49, 59, 65, 73, 83, 85, 91, 95, 103 (Constellation photographs)

Nick James – p.28 (Comet NEOWISE)

Arthur Page, Queensland, Australia – 41, 71 (Constellation photographs)

peresanz/Shutterstock – p.37 (Orion)

For the 2025 Guide, the editorial support was provided by Edward Bloomer, Senior Astronomy Manager: Digital & Data.

Further Information

Books

Bone, Neil (1993), *Observer's Handbook: Meteors*, George Philip, London & Sky Publ. Corp., Cambridge, Mass.

Chu, A (2012), *The Cambridge Photographic Moon Atlas*, Cambridge University Press, Cambridge

Dunlop, Storm (1999), *Wild Guide to the Night Sky*, HarperCollins, London

Dunlop, Storm (2012), *Practical Astronomy*, 3rd edn, Philip's, London

Dunlop, Storm, Rükl, Antonin & Tirion, Wil (2005), *Collins Atlas of the Night Sky*, HarperCollins, London

Ellyard, David & Tirion, Wil (2008), *Southern Sky Guide*, 3rd edn, Cambridge University Press, Cambridge

Heifetz, Milton & Tirion, Wil (2017), *A Walk through the Heavens*, 4th edn, Cambridge University Press, Cambridge

Heifetz, Milton & Tirion, Wil (2012), *A Walk through the Southern Sky*, 3rd edn, Cambridge University Press, Cambridge

National Geographic & Wei-Haas, Maya (2023), *Stargazer's Atlas: The Ultimate Guide to the Night Sky*, National Geographic

O'Meara, Stephen J. (2008), *Observing the Night Sky with Binoculars*, Cambridge University Press, Cambridge

Ridpath, Ian (2018), *Star Tales*, 2nd edn, Lutterworth Press, Cambridge, UK

Ridpath, Ian, ed. (2003), *Oxford Dictionary of Astronomy*, 2nd edn, Oxford University Press, Oxford

Ridpath, Ian, ed. (2004), *Norton's Star Atlas*, 20th edn, Pi Press, New York

Ridpath, Ian & Tirion, Wil (2004), *Collins Gem – Stars*, HarperCollins, London

Ridpath, Ian & Tirion, Wil (2017), *Collins Pocket Guide Stars and Planets*, 5th edn, HarperCollins, London

Ridpath, Ian & Tirion, Wil (2019), *The Monthly Sky Guide*, 10th edn, Dover Publications, New York

Rükl, Antonín (1990), *Hamlyn Atlas of the Moon*, Hamlyn, London & Astro Media Inc., Milwaukee

Rükl, Antonín (2004), *Atlas of the Moon*, Sky Publishing Corp., Cambridge, Mass.

Scagell, Robin (2000), *Philip's Stargazing with a Telescope*, George Philip, London

Sky & Telescope (2017), *Astronomy 2018*, Australian Sky & Telescope, Quasar Publishing, Georges Hall, NSW

Stimac, Valerie (2019) *Dark Skies: A Practical Guide to Astrotourism*, Lonely Planet, Franklin, TN

Tirion, Wil (2011), *Cambridge Star Atlas*, 4th edn, Cambridge University Press, Cambridge

Topalovic, Radmila & Kerss, Tom (2016), *Stargazing: Beginner's Guide to Astronomy*, HarperCollins, London

Journals

Astronomy, Astro Media Corp., 21027 Crossroads Circle, P.O. Box 1612, Waukesha, WI 53187-1612 USA. http://www.astronomy.com

Astronomy Now, Pole Star Publications, PO Box 175, Tonbridge, Kent TN10 4QX UK. http://astronomynow.com

Sky at Night Magazine, BBC publications, London. http://skyatnightmagazine.com

Sky & Telescope, Sky Publishing Corp., Cambridge, MA 02138-1200 USA. http://www.skyandtelescope.org/

Societies

British Astronomical Association, Burlington House, Piccadilly, London W1J 0DU. http://www.britastro.org/
The principal British organization for amateur astronomers (with some professional members), particularly for those interested in carrying out observational programmes. Its membership is, however, worldwide. It publishes fully refereed, scientific papers and other material in its well-regarded journal.

Federation of Astronomical Societies, Secretary: Ken Sheldon, Whitehaven, Maytree Road, Lower Moor, Pershore, Worcs. WR10 2NY. http://www.fedastro.org.uk/fas/
An organization that is able to provide contact information for local astronomical societies in the United Kingdom.

Royal Astronomical Society, Burlington House, Piccadilly, London W1J 0BQ. http://www.ras.org.uk/
The premier astronomical society, with membership primarily drawn from professionals and experienced amateurs. It has an exceptional library and is a designated centre for the retention of certain classes of astronomical data. Its publications are the standard medium for dissemination of astronomical research.

Society for Popular Astronomy, 36 Fairway, Keyworth, Nottingham NG12 5DU.
http://www.popastro.com/
A society for astronomical beginners of all ages, which concentrates on increasing members' understanding and enjoyment, but which does have some observational programmes. Its journal is entitled *Popular Astronomy*.

Software

Planetary, Stellar and Lunar Visibility (planetary and eclipse freeware): Alcyone Software, Germany.
http://www.alcyone.de
Redshift, Redshift-Live. http://www.redshift-live.com/en/
Starry Night & Starry Night Pro, Sienna Software Inc., Toronto, Canada. http://www.starrynight.com
Stellarium, https://stellarium.org/

Internet sources

There are numerous sites with information about all aspects of astronomy, and all of those have numerous links. Although many amateur sites are excellent, treat any statements and data with caution. The sites listed below offer accurate information. Please note that the URLs may change. If so, use a good search engine, such as Google, to locate the information source.

Information

Astronomical data (inc. eclipses) HM Nautical Almanac Office: http://astro.ukho.gov.uk
Auroral information Michigan Tech: http://www.geo.mtu.edu/weather/aurora/
Comets JPL Solar System Dynamics: http://ssd.jpl.nasa.gov/
American Meteor Society: http://amsmeteors.org/
Deep-sky objects Saguaro Astronomy Club Database: http://www.virtualcolony.com/sac/
Eclipses NASA Eclipse Page: http://eclipse.gsfc.nasa.gov/eclipse.html
Ice in Space (Southern Hemisphere Online Astronomy Forum): http://www.iceinspace.com.au/index.php?home
Moon (inc. Atlas) Lunar and Planetary Institute (LPI): https://www.lpi.usra.edu/resources/cla/
Planets Planetary Fact Sheets: http://nssdc.gsfc.nasa.gov/planetary/planetfact.html
Satellites (inc. International Space Station)
Heavens Above: http://www.heavens-above.com/
Visual Satellite Observer: http://www.satobs.org/
Star Chart http://www.skyandtelescope.com/observing/interactive-sky-watching-tools/interactive-sky-chart/
What's Visible
Skyhound: http://www.skyhound.com/sh/skyhound.html
Skyview Cafe: http://www.skyviewcafe.com

Institutes and Organizations

European Space Agency: http://www.esa.int/
International Dark-Sky Association: http://www.darksky.org/
RASC Dark Sky: https://rasc.ca/dark-sky-sites/
Jet Propulsion Laboratory: http://www.jpl.nasa.gov/
Lunar and Planetary Institute: http://www.lpi.usra.edu/
National Aeronautics and Space Administration: http://www.hq.nasa.gov/
Solar Data Analysis Center: http://umbra.gsfc.nasa.gov/
Space Telescope Science Institute: http://www.stsci.edu/